ANIMAL STUDIES

Prefaced with a brief introduction to the field of animal studies, the text explores the key influential terms, topics and debates which have had a major impact on the field, and that students are most likely to encounter in their animal studies classes.

Animal Studies provides a guide to key concepts in the burgeoning interdisciplinary field of animal studies, laid out in A–Z format. While Human–Animal Studies and Critical Animal Studies are the main frameworks that inform the bulk of the writings in animal studies and the key concepts discussed in the volume, other approaches such as anthrozoology and cognitive ethology are also explored. The entries in the volume attend to the differences in ongoing debates among scholars and activists, showing that what is commonly called "animal studies" is far from a unified body of work. A full bibliography of sources is included at the end of the book, along with an extensive index.

The book will be a valuable guide to undergraduate and postgraduate students in geography, philosophy, sociology, anthropology, women's studies, and other related disciplines. Seasoned researchers will find the book helpful, when researching topics outside of their specialization. Outside of academia, it will be of interest to activists, as well as professional organizations.

Matthew R. Calarco is Professor of Philosophy at California State University, Fullerton, USA.

ANIMAL STUDIES

The Key Concepts

Matthew R. Calarco

Routledge
Taylor & Francis Group

LONDON AND NEW YORK

First published 2021
by Routledge
2 Park Square, Milton Park, Abingdon, Oxon OX14 4RN

and by Routledge
52 Vanderbilt Avenue, New York, NY 10017

Routledge is an imprint of the Taylor & Francis Group, an informa business

© 2021 Matthew R. Calarco

British Library Cataloguing-in-Publication Data
A catalogue record for this book is available from the British Library

Library of Congress Cataloging-in-Publication Data
Names: Calarco, Matthew, 1972- author.
Title: Animal studies : the key concepts / Matthew R. Calarco.
Description: Abingdon, Oxon ; New York, NY : Routledge,
2021. | Includes bibliographical references and index.
Identifiers: LCCN 2020016984 (print) | LCCN 2020016985
(ebook)
Subjects: LCSH: Zoology--Encyclopedias. | Human-animal
relationships--Philosophy--Encyclopedias. | Animals
(Philosophy)--Encyclopedias. | Animal rights--Encyclopedias.
Classification: LCC B105.A55 C334 2021 (print) | LCC
B105.A55 (ebook) | DDC 590--dc23
LC record available at https://lccn.loc.gov/2020016984
LC ebook record available at https://lccn.loc.gov/2020016985

ISBN: 978-0-367-02885-5 (hbk)
ISBN: 978-0-367-02889-3 (pbk)
ISBN: 978-0-429-01932-6 (ebk)

Typeset in Bembo
by MPS Limited, Dehradun

CONTENTS

CONTENTS

INTRODUCTION

Animal Studies: The Key Concepts is intended to introduce readers to central terms, topics, and debates in the interdisciplinary field of animal studies. An ideal place to start in reading through this volume is to consider what is included under the rubric of *animal studies*. Practitioners who work in this field—ranging from professional academics and independent scholars to activists and political organizers—enter it with diverse intellectual training, political experiences, and ethical commitments; and these different backgrounds and interests have given rise to several distinct approaches to animal studies. Among the more influential frameworks in the field are Human–Animal Studies (HAS) and Critical Animal Studies (CAS); other approaches including anthrozoology and cognitive ethology also have a considerable presence.

HAS is perhaps the most established version of animal studies. HAS scholars usually trace this approach back to the pioneering works of Peter Singer (*Animal Liberation* [1975]) and Tom Regan (*The Case for Animal Rights* [1983]), both of which made a persuasive case for the centrality of animals in a variety of human practices and institutions (Shapiro and DeMello 2010). These seminal writings were followed by a host of works arguing that questions concerning animals were germane not just to philosophy (which is Singer's and Regan's home discipline) but also to geography, anthropology, sociology, women's studies, history, and related disciplines. Given that most practitioners of HAS are trained in the humanities and social sciences, the focus of their research is usually on the status and significance of animals within human societies and on the importance of the human–animal bond for understanding the formation of human individuals and culture. This emphasis on the human–animal interface does not, however, entail a lack of concern for human social justice or for research grounded in issues surrounding race, class, gender, ethnicity, and sexuality. To the contrary, HAS scholars generally insist that human–animal relations never take place entirely outside of these power-laden social relations. The point of HAS research, then, is not to displace key sociological and anthropological questions but to supplement them with analyses that demonstrate why these more standard disciplinary concerns should be seen as overlapping with human–animal relationships. (For a fuller overview of HAS, see DeMello 2012.)

CAS has a great deal in common with HAS in terms of shared historical and theoretical influences. But whereas HAS provides a space for ethico-political issues regarding animals, CAS places these concerns squarely in the foreground. CAS scholars and activists are, in general, committed to animal rights and veganism (these terms are discussed in more detail in this volume), with some proponents of CAS maintaining that veganism constitutes the moral baseline for being involved in the field. CAS, then, is driven by a more activist orientation and maintains that the primary goal of animal studies should be to contest violence and discrimination against animals. It is important to note that, according to CAS scholars and activists, the mistreatment of animals does not result simply from sloppy thinking or irrational beliefs but instead reflects the social and economic interests of the capitalist ruling class and various animal industries. Hence, the CAS framework is broadly anti-capitalist and promotes the idea that justice for animals can only be truly achieved in post-capitalist economies and ways of life. Although CAS scholars offer a variety of visions of what such alternatives might look like in practice, many proponents are influenced by a mixture of socialist and anarchist political frameworks and radical social theories such as Critical Theory (associated with the Frankfurt School) and ecofeminism. (See the introductions to Taylor and Twine 2014 and Matsuoka and Sorenson 2018 for a more extensive discussion of the CAS framework.)

The HAS and CAS frameworks inform the bulk of the writings in animal studies and the key concepts discussed in the present volume, but they are by no means exhaustive of the rich body of work that is associated with the field. There are other streams of animal studies that take their primary inspiration from areas of inquiry such as posthumanism, ethology, and environmentalism (these terms are also discussed in more detail in the entries that follow). In view of these disparate lines of analysis and perspective, some scholars have argued that there is little point in trying to synthesize such a complex and often incompatible body of research into a single field with the name "animal studies." Rather, they suggest, we should accept the differences among these approaches and allow for other rubrics and nomenclatures to emerge (see the introduction to Lundblad 2017 for a powerful version of this argument). Following in this vein, there are scholars and activists who argue for the relative distinctiveness and autonomy of such fields as animality studies, anthrozoology, animal ethics, cognitive ethology, philosophical ethology, and so on. The point here is not to adjudicate these nomenclature disputes or recommend one approach over another; it is sufficient for readers new to the field simply to be aware of the fact that what is commonly called "animal studies" is far from a unified body of work and that there are ongoing debates among scholars and activists about precisely what is at stake in such work. The entries in the volume will attend to these differences where appropriate but also mark points of overlap where they exist.

Today, courses in animal studies are offered in several disciplines across the arts, humanities, and social sciences, including anthropology, cultural studies, environmental studies, film, geography, history, literature, philosophy, sociology, and many others. Given the constraints of space, I have chosen to focus primarily on key concepts rather than providing overviews of how animal studies is practiced in these different disciplines. However, if you are taking a course in animal studies attached to a specific disciplinary orientation, I highly recommend exploring a book like Margo DeMello's *Teaching the Animal* (2010) in order to get a better sense of how your particular discipline is generally oriented toward the field.

Finally, I offer a word about the criteria for selection of the key concepts included here. No book can be all things to all people, and I am keenly aware that the present work will fail to include concepts, texts, and discussions that some advanced scholars consider essential in animal studies. That said, the list of key concepts covered in this volume is neither arbitrary nor idiosyncratic. It was generated from feedback from nearly a dozen leading scholars and established instructors in the field as well as four anonymous readers for the press (feedback and suggestions for which I am deeply grateful). The list also derives from my own experience with teaching animal studies courses in a variety of settings for nearly two decades. Thus, even though this volume is not intended to discuss every key concept that might arise in such courses, my sincere hope is that several of the entries will be of use to students doing research for their animal studies courses as well as for readers and researchers who hope to gain a better sense of the lay of the land in the field of animal studies.

ABOLITIONISM

Abolitionism is a moral, legal, and political approach that entirely rejects all forms of animal exploitation. The model for this perspective derives from the human rights paradigm in which human beings are understood to have a basic moral right to be treated as ends in themselves rather than as mere means to another person's end (that is, non-instrumentally). In the legal and political realm, the human rights approach entails granting human beings strong protections against their rights being violated and abolishing institutions that violate such rights. Seeing human beings as having basic rights, it is argued, is what provides the grounding for the complete abolition of human slavery and other forms of non-consensual violence. Correspondingly, if animals have similar rights and protections, then we ought not simply *limit* their exploitation; rather, we should, according to this line of reasoning, *abolish* it entirely.

Of course, the persuasiveness of this position relies on there being deep similarities between human beings and animals that can justify their similar treatment. Abolitionists view human beings and animals as essentially the same at level of *moral* consideration—which is to say, that despite certain differences among humans and animals on certain registers (for example, cognitive, physiological, and social), at a moral level animals and humans are equally worthy of full moral consideration inasmuch as they are both sentient entities with preferences, desires, and interests. If an animal is sentient and has its own interests, abolitionists argue that those interests ought to be taken into account, thereby ruling out a purely instrumental attitude.

In view of practices such as **meat** eating, where an animal's interest in living and avoiding pain are very high and where the consumer's interest in having their taste preferences satisfied are comparatively low, the abolitionist position is particularly strong. The meat industry could, in principle, be eliminated in practice and prohibited at the legal and political levels in a fully functioning society. With regard to more controversial issues such as animal **experimentation** for non-trivial purposes (for example, research on life-saving medicines and procedures), the abolitionist approach would have us consider whether it would be justifiable to institute non-consensual experimentation on human beings in similar situations. If not, then we ought as a matter of logical consistency rule out any such experimentation on animals.

One of the key features of the abolitionist position is its uncompromising rejection of **animal welfare** practices, strategies, and campaigns. Inasmuch as animal welfare organizations seek merely to improve the lives of exploited animals rather than eliminate such exploitation altogether, the abolitionist position recommends eschewing welfarist logic. Recently, many organizations that purportedly support animal rights have argued that using welfare reforms in the short term is an effective part of a long-term strategy for achieving animal rights. Dubbed "new welfarism" (Francione 1996), abolitionists also reject this approach, arguing that welfare reforms do not lead to long term gains for animal rights and that they ultimately provide justifications for ongoing, more "humane" forms of animal exploitation. By contrast with reformism of both sorts, abolitionists tend to argue that particular exploitative practices should be eliminated where possible and in an incremental fashion. Given that the highest number of animals are killed and exploited in **CAFO**s, abolitionists have focused much of their energy on vegan outreach and peaceful legal and political advocacy against meat eating, hoping to abolish this practice entirely before moving on and challenging other exploitative industries and practices. Critics of abolitionism have suggested that it is unable to provide a groundwork for positive human–animal relationships (Donaldson and Kymlicka 2011); others have suggested that abolitionism tends to underestimate the persistence of racism and sexism and various forms of exploitation in societies that are typically thought to be abolitionist (Kim 2018).

Further reading: Francione 1996; Francione 2008; Francione and Garner 2010; Nibert 2013; Steiner 2008; Wrenn 2015

ABSENT REFERENT

A term developed by Carol Adams (2015), the absent referent describes the structures and processes that deprive individual beings (that is, referents) of their unique subjectivity (that is, their living presence). Animals become absent referents chiefly in the production of meat for human consumption. By the time most consumers encounter animal flesh in the grocery store or on their plates, the individual animal that gave its life for that meat has been rendered effectively absent. Thus, instead of interacting with living, present animals who have unique personalities and subjectivities, industrial processes of meat production and consumption present us with already butchered, dismembered chunks of flesh, renamed as "steak," "hamburger," "veal," "pork chop," and so on. According to Adams, this dual process of butchering and re-naming is at the heart of turning living animals into absent referents and forms the core of the structural and practical means whereby animals come to be removed from ethical and political consideration in advanced industrial societies.

Animals that have been turned into absent referents in this literal way are also rendered absent referents in more indirect, metaphorical ways. Thus, one sometimes says: "I feel like a piece of meat;" but such statements fail fully to appreciate the ways in which meat is actually produced. Further, one cannot really feel like a piece of meat and still utter such statements, since to be rendered into consumable meat, one's ability to express oneself in this kind of way (that is, through **language**, whether human or nonhuman) is concomitantly eliminated.

Although the concept of the absent referent is meant primarily to explain how animals are deprived of their singular living presence in meat eating and other similar commodified practices, Adams argues that the concept is also at work in interhuman relations, for example, in regard to gender and racial discrimination. With regard to gender in particular, Adams suggests that in patriarchal–industrial cultures women and animals should be understood as *overlapping absent referents*. At issue here is the idea that women, too, are sometimes deprived of their uniqueness, singularity, and subjectivity by dominant systems and ideologies, and thus suffer similar (but not identical) forms of violence. Through ideological means (for example, media, pornography, marketing, and so on), women's bodies are fragmented and objectified, focusing the viewer on women's body parts rather than on the integrated, individual woman.

Many times, such processes of objectifying and fragmenting women borrow heavily from language and imagery used to describe and denigrate animals. Women's sexualized body parts are described using animal terms, and individual women are sometimes classified alongside animals. Adams describes how these processes form a kind of feedback loop that rebounds back on animals, leading in turn to animals being sexualized and feminized. In the marketing and advertising of animal flesh and byproducts, animals are sometimes portrayed as desiring their consumption or seducing the consumer to eat them. Adams summarizes this point by suggesting that "in a patriarchal, meat-eating world animals are feminized and sexualized; women are animalized" (Adams 2016: 165).

The concept of the absent referent also implies that some beings *are* allowed to maintain their "presence," which is to say, their individuality and integrity. In patriarchal, meat-eating societies, the quintessential example of such a subject is the human male. Human males are seen as full subjects who interact with various kinds of objects that fall short of the status of being full subjects. Insofar as both animals and women belong to the class of beings who are typically denied full subjectivity, there are good reasons for considering and contesting their oppression jointly rather than separately. Thus, for Adams, feminists should be informed and supportive of arguments in defense of animals, and those who support animal well-being should be interested in feminist politics and feminist analyses of oppression.

The links posited by Adams between the oppression of animals and women are not meant to indicate some deep, essential bond between femininity and animality, as if women are somehow closer to animals by nature. Rather, the overlapping nature of the oppression of women and animals is a contingent, historical observation about the nature of exploitation in contemporary, patriarchal, meat-eating societies. Thus, the concept of the absent referent is not meant to capture a timeless fact about human cultures but rather aims to un-cover a certain historical logic of oppression that sub-humanizes animals and women (and other marginalized groups) in distinct but overlapping ways. Recently, critics of the concept of the absent re-ferent who are otherwise supportive of both feminism and animal defense, and who acknowledge connections between the exploitation of women and animals, have called for more careful and refined analyses of how pornographic representations of women and wo-men's sex work are intertwined with violence against animals (Grebowicz 2010; Hamilton 2016) as well as fuller consideration of how racism and **colonialism** are at work in the production of absent referents (Deckha 2012).

See also: CAFO; carnophallogocentrism; dehumanization; ethics; feminism; intersectionality; race

Further reading: Adams 2004; Adams 2016

ACTIVISM

Activism on behalf of animals is aimed at changing their ethical, social, legal, and economic status. Pro-animal advocacy of this sort extends back in time to a number of ancient cultures and traditions.

Wherever the consumption, sacrifice, and instrumental use of animals has been practiced historically, there have been notable counter-movements calling for the reform and even abolition of such practices (Phelps 2007). In modern times, pro-animal activism has become more widespread and more organized. Today, there are dozens of national and international organizations dedicated to **animal welfare and animal rights**, and grassroots activists have a visible presence in all major societies and nations.

Historians date the origins of this upswing in animal activism and advocacy to the 17th century, when some of the earliest efforts to pass laws aimed at addressing animal cruelty were proposed. This legislation was followed soon thereafter by the establishment of organizations aimed at ensuring animal welfare, with the founding of the Royal Society for the Prevention of Cruelty to Animals in England in 1824 and the American Society for the Prevention of Cruelty to Animals in 1866. Organized movements and societies for vegetarianism also emerged during this time, often with a focus on the health benefits of a vegetarian diet but sometimes in view of animal welfare and curbing the violence of animal slaughter. The end of the 19th century saw the publication of Henry S. Salt's *Animals' Rights Considered in Relation to Social Progress* (1894), which summed up and extended a number of pro-animal advocacy works from antiquity to Salt's own era (see Finsen and Finsen 1994 for a helpful overview of this history).

It was in the middle-to-late 20th century that the modern era of animal activism emerged. During this period, the first explicitly vegan societies were formed and a number of books documenting and criticizing the horrors of factory farming and animal experimentation were published. In the 1970s and 1980s, Peter Singer's *Animal Liberation* (1975) and Tom Regan's *The Case for Animal Rights* (1983) appeared and gave voice to the aims of a number of animal liberation and animal rights activists. Also in these years, some of the most prominent and influential pro-animal organizations were founded, including People for the Ethical Treatment of Animals, In Defense of Animals, Animal Legal Defense Fund, and Farm Sanctuary. Alongside these more mainstream approaches, a more militant form of animal rights activism emerged, associated with underground activist cells.

As this cursory sketch indicates, modern animal activism comprises a wide variety of organizational and outreach approaches, each of which is premised on different strategies and tactics. Philosophers and intellectuals associated with these movements have tended to concentrate on transforming the beliefs of individuals through education; and, in some cases, they aim to supply or articulate the rationale behind pro-animal activism. The leading idea here is that animal welfare and animal rights, which have long been mocked for being irrational and sentimental, do in fact rest on well-grounded and cogently reasoned arguments; if the public is exposed to these arguments, the thinking goes, it will have a transformative effect on the liberation of animals.

Mainstream pro-animal organizations, by contrast, have focused chiefly on creating highly visible campaigns to raise awareness about extreme acts of animal abuse. The hope is that these awareness-raising campaigns will lead to public outrage and consumer boycotts, which will in turn force animal-abusing industries either to shut down or reform their practices. This approach has led to a variety of partial and full victories for pro-animal activists. For example, through such campaigns some companies and industries have ceased animal testing, adopted different slaughtering practices, and freed confined animals. Supplementing these efforts, animal lawyers use the tools of **law** and public policy to prevent animal abuses and transform industries. Animal lawyers are involved in a range of practices, from ensuring that animal cruelty laws are enforced to arguing for the extension of full legal personhood to certain animals. These large-scale campaigns and legal initiatives have been supported and sometimes generated at the local level by grassroots activists who promote animal liberation and welfare through vegan and vegetarian outreach, caring for abandoned animals at **shelters and sanctuaries**, and advocating for pro-animal policies in their communities and states (see Guither 1998 for a fuller overview of traditional activism).

As laudable as such mainstream activism is, there is a sense among activists of all stripes that progress on behalf of animal liberation has been disappointingly slow-going; moreover, for every advance made by activists in one area, violence against animals seems to increase in another. Frustration over the failure of mainstream animal activism to bring about sweeping changes in society has led to the emergence of the direct-action, militant approaches mentioned above. Direct-action activists hope to inflict severe economic damage on animal-abusing industries (for example, by ruining laboratory equipment and research materials used in animal experimentation), with the aim of making their practices too expensive and burdensome to continue. The guiding idea behind much of direct-action activism is that State apparatuses (for example, law, education, and electoral politics) are unable and unwilling genuinely to transform social relations with animals. Direct-action activists argue that the primary aim of the State, especially in advanced capitalist societies, is to create institutions that ensure the smooth functioning of **capitalism** and protect the interests of wealthy and powerful human beings. Lobbying this system to ensure the rights and welfare of animals thus strikes most direct-action activists as a waste of energy that would be better spent in engaging in actions that make an immediate difference in animals' lives (see Best 2014 and Best and Nocella 2004 for further reflections on this approach to animal activism).

Even with all of these approaches held simultaneously in view, a snapshot of the current state of animal activism provides little grounds for optimism. The movement has achieved only minor successes, and the exploitation of animals is increasing in nearly every register. The relatively small size and limited successes of the animal liberation movement have recently led many long-time activists to consider ways to broaden the movement's influence (Cavalieri 2016), including the creation of coalitions with various constituencies (such as progressive social and ecological movements) as well as the radical transformation of the political sphere (for example, through extending citizenship and other political rights to animals).

See also: abolitionism; intersectionality; total liberation; veganism and vegetarianism; zoopolis

Further reading: Finsen and Finsen 1994; Phelps 2007

AGENCY

Agency is defined as the ability to act, make choices, and exert power. The term is best applied to actions that are *constrained* by certain factors but not fully *determined* by them. Traditionally, agency has been thought to be unique to human beings and has played a central role in the establishment of classical versions of the **human/animal distinction**. Many animal studies theorists, however, argue that some animals are also agents and are capable of acting in ways that cannot be fully explained in terms of determined, unconscious instincts and drives. Such work on animal agency belongs to a broader trend within the field of seeing animals not simply as inert, mechanical objects but as subjects who negotiate a complex field of possibilities within their environments.

When considering animal agency, researchers have a range of animal actions in mind, including such things as: refusing to **work**; fighting back against abusive treatment; using novel means to escape sites of **captivity**; reestablishing territory that has been disrupted by development; and organized forms of group resistance. Recent work by critical animal studies theorists and animal biologists has unearthed a whole host of astonishing anecdotes and events that illustrate and explain such actions. For example, critical animal studies historian Jason Hribal (2010) offers compelling accounts of numerous acts of animal resistance in circuses, aquariums, and zoos, many of which involve novel uses of tools, strategies, and forms of cooperation between animals that would seem to imply extraordinarily high levels of agency at the individual and collective level. Some ethologists and philosophers have gone a step further to suggest that animals also exhibit a sense of justice and can be considered moral agents in their own right, not just the passive recipients of moral consideration by human beings (Bekoff and Pierce 2009; de Waal 2006; Rowlands 2012).

Scientists and theorists who are skeptical of animal agency are sometimes quick to reject these kinds of accounts and argue instead that the sorts of actions just mentioned can be explained by unconscious, instinctual behaviors. But many animal studies theorists (including biologists and ethologists who work carefully with the animals under discussion) believe that the evidence for animal agency is overwhelming at this point. The question, they suggest, is less a matter of whether there is animal agency and more a matter of what kinds of animal agency exist. So, rather than insisting on a wholesale rejection or affirmation of animal agency, we might instead see agency as running along a continuum from (a) the minimal sense of animals having some effect on history and the world, to (b) more complex individual acts of choice or decision-making, to (c) the maximal sense of animals engaging in fully conscious, intentional agency in collective and organized forms. While most of the recent literature in animal studies seems to accept (a) and (b) as reasonable ways of thinking about animal agency, there is considerable debate about whether animal agency rises to the level of (c) (see Carter and Charles 2013 and Pearson 2015 for arguments against more advanced forms of animal agency). For those advocates who believe that animals *do* exhibit this kind of sophisticated, intentional form of agency and resistance, pro-animal **activism** is sometimes seen as going beyond the traditional paternalistic model (where human activists serve as a "voice for the voiceless" animals, who are thought to need human representatives to express their interests [Suen 2015]) to a more complex solidarity model (in which human activists aid animals' own resistance and seek to clear away obstacles that allow animals to pursue their agential lives more freely [Coulter 2016]).

See also: anthropomorphism; world

Further reading: Massumi 2014; McFarland and Hediger 2009; Rasanen and Syrjamaa 2017; Steward 2009

ANIMAL–INDUSTRIAL COMPLEX

First coined by anthropologist Barbara Noske (1989), this concept refers to the wide variety of socio-economic forces at work in the exploitation of animals. Scholars who employ this term maintain that **capitalism** and the profit motive in particular are largely, but not exclusively, responsible for the widespread violence directed toward animals. The concept functions on partial analogy with other "industrial complexes," such as the military–industrial complex, the prison–industrial complex, and the pharmaceutical–industrial complex. In all of these complexes, there exists a tight, interconnected relationship between a particular social sector and economic interests such that the one cannot be fully understood without reference to the other. If we were to apply this "industrial complex" perspective to the consumption of animals, for example, we would notice that it is not primarily consumer preference for animal flesh that explains the extraordinarily large numbers of animals killed and eaten, but rather the shared economic interest that various animal industries (farmers, slaughterhouses, grocers, and so on) have in continually increasing the amount of meat consumed (see **CAFO**). Recently, the term has been used more expansively to denote not just the economics of animal exploitation but also the wide range of affiliated institutions, technologies, affects, and bodies of knowledge that are linked with and sustain animal industries (Twine 2012).

Further reading: Nibert 2013; Stallwood 2013

ANIMAL LIBERATION

Animal liberation is a concept used in a wide variety of contexts, but it is most closely associated with the writings of philosopher Peter

Singer and his 1975 book *Animal Liberation*. This entry first explores how the term is used in Singer's work and then briefly discusses how the term circulates in related contexts. For Singer, animal liberation is the most recent in a series of liberation movements. Writing in the 1970s, when movements for Black, gay, and women's liberation were gaining global recognition and broad support, Singer argued that there was no persuasive reason for why animal liberation should not be added to the list of progressive political struggles. While struggles against racism, homophobia, and sexism include only human beings within their scope, Singer argued that recognizing the biological humanity of people who are discriminated against is not what is truly at issue in these movements for justice. Rather, what is at stake is recognizing that racism, sexism, homophobia, and other such forms of discrimination are problematic because they override the interests and preferences of marginalized individuals without good reason for doing so. In other words, what makes these kinds of discrimination problematic is that they treat marginalized human individuals as if their preferences, interests, and desires either do not matter or matter less than those of the dominant group. Liberation movements, according to Singer, demand that the interests of *all* individuals be given **equal consideration**. There is no ethical justification for ignoring an individual's interests; the fact that someone is a different gender, has a different sexual preference, or belongs to a different ethnic group does not mean they do not have interests of their own. And to protect one's own (or one's own in-group's) interests while failing to consider others' interests is, according to Singer, the kind of partial treatment that ethical reasoning demands we challenge.

But, one might wonder, how do these ideas about interests and equal consideration apply to animals? For Singer, the link between human and animal liberation lies in the shared "logic" of liberation. As just noted, human liberation is not so much about liberating human beings because they belong to the species *Homo sapiens*; rather, humans should be liberated from unjust treatment because they prefer and have an interest in being treated fairly. The same holds true, Singer argues, for animals. Animals are not like human beings in every conceivable way, but they are like human beings in being sentient (see entry on **sentience**) and thereby having interests and preferences. Although we may not be able to discern every interest or preference a given animal might have, it is evident that animals resist such things as being slaughtered, undergoing painful experimentation, and similar forms of harmful treatment. As such, the logic of liberation asks us to consider whether such actions can be justified and to give full and equal consideration to the interests of animals in so doing. To act otherwise and be partial to one's own species without sufficient justification is to be guilty of **speciesism**, according to Singer. On Singer's account, there *are* instances where we might be able to rationally justify killing and eating an animal or experimenting on it, but the standard ways in which animals are raised and slaughtered in **CAFO**s or experimented on in scientific settings fail to meet those rational standards. Consequently, Singer's version of animal liberation leads to radically curtailing most forms of animal consumption and **experimentation**.

Singer's *Animal Liberation* has been enormously influential in both animal **ethics** and animal studies as well as among mainstream animal activist organizations. His book is often credited with being the "Bible" of the animal liberation movement and has played an important role in helping to define the mission of several prominent **animal welfare and animal rights** organizations. At the same time, Singer has been subjected to thoroughgoing critique by animal rights thinkers and abolitionists who claim that his specific ethical approach permits various kinds of animal abuse and exploitation (see the entry on **abolitionism**). Singer has also been challenged by feminists for the excessive rationalism of his work (see the entry on **feminism**) and by disability theorists for his insensitivity to the comparisons he draws between so-called "marginal case" human beings and animals (see the entry on **disability**).

The term animal liberation is also used to refer to the animal defense movement as a whole, especially in regard to those actions and campaigns that aim to liberate animals from **captivity**, caging, and confinement. These kinds of animal liberations are often carried out anonymously by activists who conceal themselves behind masks (for example, with Animal Liberation Front activists) as well as by activists who do open rescues, identifying themselves either by video or in writing and who risk arrest in the process (see the entry on **activism**).

Further reading: Jamieson 1999; Newkirk 2000; Schaler 2009; Singer 1993; Singer 2001a

ANIMAL WELFARE AND ANIMAL RIGHTS

Animal welfare and animal rights are two related but distinct ways of conceiving of the ethical, political, and legal standing of animals within human societies. This entry begins with a discussion of animal welfare and then briefly examines the ways in which the animal rights approach both overlaps with and differs from animal welfare.

The term animal welfare is usually used to refer to the well-being of animals, and is understood variously to involve an animal's pleasure, state of happiness, ability to flourish, and freedom to live out a natural way of life in its natural environment. Animal welfare advocates seek to ensure any one or a combination of these aspects of well-being for individual animals or species as a whole. Although concern with animal welfare is sometimes thought to be a relatively recent phenomenon, there are in fact a number of Eastern, Western, and Indigenous cultural, religious, and philosophical traditions that grant central importance to animal welfare. Many of these traditions inform the modern animal welfare movement, which comprises a number of prominent intellectuals, activists, and national and international organizations.

Given the diversity of approaches to animal welfare, it is impossible to offer a one-size-fits-all definition that will capture its essence. However, one recurrent feature of animal welfare organizations and philosophies is a commitment to reducing cruelty and violence toward animals on the one hand and increasing humane and kind treatment on the other. While these broad aims are consistent with the abolition of certain violent practices and industries, most animal welfare organizations and activists tend to take a more moderate, reformist approach to human–animal interactions. For most animal welfarists, the industrialized use of animals is not inherently unethical; rather, the problem with modern animal industries is their often unnecessarily cruel and disrespectful practices and policies. In line with this perspective, animal welfarists typically seek to ameliorate the worst forms of violence in various institutions by proposing reasonable and pragmatic reforms to existing practices and policies. To illustrate, in recent years animal welfare organizations and activists have advocated for changes in the ways in which food animals are treated during the rearing and slaughtering processes, making concrete suggestions for better living conditions and slaughtering methods. Similarly, animal welfarists have put their energies into supporting the adoption of abandoned or unwanted **pets** and seeking stricter prosecution of domestic animal abusers. Another common focus of animal welfarists is animal rescue, which focuses on saving both domestic and wild animals who have been abandoned or dislocated during natural disasters and similar emergency situations. Animal welfarists do, however, tend to believe that certain practices involving animals are inherently cruel and should be abolished altogether. So, unnecessary practices such as trade in exotic animals, cutting the fins off of sharks, trophy **hunting**, animal racing, animal fighting for entertainment, and related forms of cruel treatment would, from the welfarist perspective, have no place in a way of life or culture dedicated to animal well-being.

Animal rights activists, theorists, and organizations share with welfarists the idea that these sorts of cruel practices should in principle be abolished. But they go further than welfarists in arguing that nearly *all* institutionalized human uses of animals should be abolished or radically altered. The animating idea behind the animal rights position is that animals are *subjects-of-a-life*, which is to say, conscious subjects who have a stake in how they are treated and whether they live or die (Regan 1983). Hence, animal rights advocates believe that animal subjects, like human subjects, are rights-bearing individuals who ought never to be treated in a merely instrumental manner—that is, as a mere means to some end. Given that many human–animal relations take this instrumental form (for example, with the institutionalized use of animals for their meat and fur, with the exploitation of animals for **entertainment**, testing, and so on), it follows from the animal rights perspective that they should be entirely abolished, not just reformed. (For further discussion of the ethics and politics of animal rights, see the entries on **ethics** and **abolitionism**.)

See also: activism

Further reading: Cavalieri 2001; Grandin 2015; Keeling, Rushen, and Duncan 2011; Palmer and Sandøe 2018; Regan 2001; Singer 2001a; Wise 2000

ANIMOT

Animot is a neologism invented by Jacques Derrida (2008) to emphasize the rich multiplicity of animal life, as well as the complex and ambiguous nature of animals' relationship to **language**. Before breaking this term into its component parts and exploring its specifically French linguistic character, it will be helpful to contextualize the term within Derrida's broader approach to what he calls "the question of the animal."

Derrida argues that common ways of thinking about animals, which have been inherited in large part from the Western philosophical and intellectual traditions, are inadequate and reductive. We have been encouraged to think of animals as constituting one large set of beings who share a common essence. We have been similarly led to believe that, on the other side of that set, there is a single, homogeneous humanity that can be definitively separated from animal existence. This classical conception of the **human/animal distinction** would thus have us oppose The Human to The Animal. Such binary and oppositional thinking, according to Derrida, fails to do justice to the complexities of the regions of existence under discussion. To reduce the immense diversity of animal life to a category such as The Animal is to fail to attend to the ways in which animal life exceeds any single, overarching definition. Similarly, a reductive conception of animals leads us to think of what is supposed to be animals' binary opposite (in this case, The Human) in the same kind of reductive terms. Thus, rather than appreciating the rich diversity we see among human cultures and the singularity of individual human beings, we tend to focus only on some supposedly universal human essence (rationality, the capacity for language, or awareness of **death**, and so on).

In Derrida's analysis of the question of the animal, this conceptual violence and reductionism is at the heart of the physical and institutional violence visited on animals within Western cultures. Placing singular animals and the rich diversity of species within conceptual cages is the first step on the path toward literally capturing and caging animals and subjecting them to untold forms of violence. In order to develop a more generous and respectful relationship with animals, Derrida believes we must develop other ways of speaking and thinking about them as well. His invention of the term *animot* is a contribution to that transformational project.

Derrida uses the term *animot* in place of The Animal (*l'animal* in French). In replacing the essentialist term The Animal with *animot*, Derrida hopes to accomplish three things. (1) In line with the points about reductionism just mentioned, he wants to recall his readers to the plurality and rich multiplicity of animal life. When spoken in French, *animot* sounds like the plural word for animals, *(les) animaux*. So, rather than speaking of a general animal or animal-ness, the word *animot* reminds us that animals are plural and irreducibly unique.

The *mot* (which means *word*) in *animot* refers to two additional aspects of the question of the animal that Derrida aims to bring to our attention. (2) With a focus on the word, Derrida calls on readers to reconsider how language and the ability to see and name things have been used throughout the tradition to differentiate human beings from animals. Animals have often been characterized as entirely lacking in language and human beings as universally having the capacity for speech. But, for Derrida, having a language and using words are a more complex matter than that. Language is by no means a transparent tool for gaining access to reality, and it is not fully within human control (we receive it from others as we grow up within and become a competent member of a speaking culture). Conversely, animals do not simply "lack" language, nor are they unresponsive automatons who react in a simple and mechanical way to their environment. (3) With this more complex conception of language in place, Derrida does not go on simply to suggest that human beings and animals have the exact same relationship to language. In other words, he does not seek to "give language back" to animals; rather, he tries to think about animals existing outside of human language in a way that does not imply privation or lack on the part of animals.

Further reading: Krell 2013; Lawlor 2007; Naas 2015; Westling 2014

ANTHROPOCENE

Anthropocene is a term that was initially suggested by scientists Paul Crutzen and Eugene Stoermer (2000) to name the recent geological era in which human beings have become the main drivers of planetary change. The origins of this era can be dated, they argue, to roughly two hundred years ago, when a variety of human activities began to have a noticeable and measurable influence on planetary systems. This starting point also coincides with the Industrial Revolution, a period that ushered in extensive economic and technological innovations. Among the measurable impacts that human beings have had on the planet during this period, the most consequential include: exponential human population growth, coupled with increased resource use and urban sprawl; massive land surface changes due to development and agriculture; an exponential increase in the rate of species **extinction**; the use of more than half of the earth's accessible fresh water; and increased emissions of greenhouse gases that have contributed to global **climate change**. Other scientists and researchers argue that the Anthropocene should be dated to some 8–10,000 years ago, when large-scale agriculture and animal domestication began and human beings first markedly altered their natural environment (Ruddiman 2003). Whatever historical marker is used, the Anthropocene hypothesis proposes that human beings have, in the more-or-less recent past, transformed the geology and ecology of the earth in an epochal manner.

Scholars in animal studies have had mixed responses to the Anthropocene hypothesis, with some who strongly support and argue for its usefulness and others who maintain that the term fails to capture the scale and nature of the planetary problems facing human beings and animals. Political theorist Wayne Gabardi (2017) finds the term helpful for highlighting how the socio-economic forces of late modernity have sought completely to colonize and dominate the earth. With regard to animals in particular, he maintains that the Anthropocene hypothesis can be used to pinpoint the ways in which anthropogenic (or, human-caused) effects on ecosystems and natural habitats threaten to obliterate huge numbers of individual animals and entire animal species, possibly bringing about a sixth mass extinction. By remaining with the strong version of the Anthropocene hypothesis, Gabardi believes we might finally come to grips with how ecologically and zoologically devastating modern lifestyles are. Further, we might come to recognize that modernity and its anthropocentric social contracts are not up to the task of dealing with the ecological devastation that has been wrought. In order to address the planetary issues that characterize the Anthropocene, Gabardi argues that our old anthropocentric social contract must give way to a **more-than-human** social contract that is constructed in view of the well-being of both human and animal life.

Morten Tønnessen and Kristin Armstrong Oma (2016) also believe the term Anthropocene can be useful for animal studies. They suggest that, although the term has been used to date in primarily anthropocentric ways, an increased focus on the status of animals within the Anthropocene can deepen the framework of animal studies and enrich ongoing discourse surrounding the Anthropocene. They further suggest that by considering the devastating effects of projects of human domination during the Anthropocene, the need to reverse course and give up the goal of human dominion might finally become evident. Jamie Lorimer (2015) makes a similar argument about the importance of the notion of the Anthropocene from the viewpoint of **wildlife** conservation, arguing that the concept allows us to think more carefully about the deep entanglement of the human and more-than-human worlds, thereby helping to overcome the previous paradigm that views wildlife as something "out there" and largely separate from human affairs.

Critics of the Anthropocene hypothesis have challenged its usefulness in several ways. Andreas Malm and Alf Hornborg (2014) voice a common concern about the failure of the hypothesis to mark salient subdivisions among human beings in relation to the main drivers of ecological degradation. Rather than "humankind" as such being responsible for the advent of contemporary ecological degradation and climate change, they argue that only a very small percentage of human beings has caused these problems. The vast majority of human beings, they maintain, have neither ruined the ecology nor benefitted from the socio-economic systems that produced such ruin. This argument implies, conversely, that a number of cultures and ways of life have existed before and during the Anthropocene that do not lead to the mass extinction of animal species or to widespread ecological collapse. Thus, the current devastation of animals and the more-than-human world cannot be understood simply as something produced by the human species as a whole.

Animal and environmental studies theorist Eileen Crist (2013b) suggests that the Anthropocene hypothesis is being used to extend the very technological and managerial mindset that gave rise to the current predicament. If we remain within this dominant paradigm, Crist argues, we will continue to believe that the solution to the problems associated with the Anthropocene hypothesis lies in deepening our dominion over animals and the rest of the earth to ensure the sustainability of present human lifestyles. For Crist, by contrast, the scale and scope of the harm done to animals and the earth calls for a radical challenge to the modern worldview and current ways of life. She argues that, rather than trying to increase our dominion over animals and the earth, we need fundamentally to pull back our influence, reduce our ecological footprint, and make additional room for more-than-human species to flourish.

In line with these concerns, Donna Haraway (2016b) also worries that the concept of the Anthropocene misleads us into thinking that the human species is a unique, stand-alone species that can generate technological solutions to the problems it has created. Avoiding both technological optimism and fatalistic despair in regard to current ecological issues, Haraway maintains that we need to learn to "stay with the trouble," that is, to continue to see ourselves as **companion species** with animals and other earthly beings and to work with and in view of their well-being as we co-create future worlds with them.

She recommends leaving behind the Anthropocene era and trying instead to inaugurate a new era that she dubs the Chthulucene (that is, an era focused on the well-being of the earth [*chthon*] as a whole). For Haraway, the overarching goal of this era would be to root ourselves in the earth ever more deeply and to affirm respectful relations with earth others. Similar to Crist, Haraway argues that the overarching ethical task is to learn to make room for animals and the rest of the more-than-human world and to co-create the conditions for the joint flourishing of the human species and species of all sorts.

See also: climate change; extinction

Further reading: Human Animal Research Network Editorial Collective 2015; Tønnessen, Oma, and Rattasepp 2016

ANTHROPOCENTRISM

Anthropocentrism is the view that human beings (in opposition to animals and other nonhuman beings) are of supreme importance in ethical, political, legal, and existential matters. While anthropocentrism can be found in a number of cultures, it plays a particularly strong role in the Western tradition, featuring prominently in texts and practices that extend back historically as far as ancient Greece (Renehan 1981; Rodman 1980). The concept has a wide circulation in animal studies, environmental humanities, and **environmentalism**, where it has generally been subjected to strong criticism. Among the primary characteristics of anthropocentrism are: (1) a narcissistic focus on human exceptionalism; (2) a binary account of human–animal differences; (3) a strong moral hierarchy that ranks human beings over animals and other nonhuman beings; (4) a tendency to de- and sub-humanize certain populations; and (5) institutions that aim to protect and give privilege to beings deemed fully human. This entry examines each of these characteristics in turn.

Anthropocentrism should be seen primarily as a kind of incessant attention to and rotation around exclusively human existence. Recurrent efforts are made in anthropocentric discourses to demonstrate the exceptional and remarkable status of human beings, a gesture that Sigmund Freud describes as *human narcissism* (Freud 1955). One consequence of this kind of human narcissism is that, when efforts *are* undertaken to examine the lives of nonhuman others, such inquiry tends to be carried out primarily in view of human perspectives and projects, reinforcing an instrumentalist view of the nonhuman world.

In regard to animals in particular, anthropocentrism maintains that human beings and animals can be sharply distinguished and differentiated along various lines. These distinctions are most often figured through a series of traits that purportedly belong solely to human beings and are said to be "lacking" in animals, as though animals are in some way deficient or impoverished in comparison to human beings. The traits or capacities that animals have been thought to lack include rationality, **language, mind**, and **agency**, among others. Even in a post-Darwinian age, when upholding such oppositional distinctions has become increasingly untenable from a biological and evolutionary perspective, it is still common in many (academic and non-academic) contexts to encounter an anthropocentric insistence upon human uniqueness over and against animals (for more on this point, see the entries for **evolutionary theory** and **human/animal distinction**).

Another recurrent tendency of anthropocentrism is to couple these kinds of oppositional distinctions with a moral hierarchy in which human beings are given higher value than animals and other non-human beings. One of the problems with this kind of normative ranking is that there is often no compelling reason given for *why* animals who might lack a specific capacity or ability should have less ethical value. For example, simply because a particular animal might not be able to make moral decisions does not necessarily or logically entail that such a being should not have as much ethical standing as a human being who does.

Anthropocentrism is not, however, limited simply to conceptually separating and giving ethical privilege to human beings over animals and other nonhuman beings. Paradoxically, anthropocentrism also creates divisions within and among human beings themselves. In other words, anthropocentric attitudes have not historically served to grant the status of full humanity to all biological human beings but only to a select subset of the species—namely, those who are deemed by the dominant culture to be "truly," or "fully," or "properly" human. This kind of delimitation of what counts as properly human plays an important role in the establishment of sexism, racism, and similar kinds of **dehumanization** (for more on this aspect of anthropocentrism, see the entry for **anthropological machine**).

Finally, it is important to stress that anthropocentrism is more than just a conceptual apparatus. Anthropocentric ideas, attitudes, and dispositions are enacted through a robust and interlocking series of practices and institutions. Primary among these institutions is the **animal–industrial complex**, which exploits animals for food, for their byproducts, and for their labor power; anthropocentrism is also evident in the military, medical, and pharmaceutical industries that use animals for experimental and biotechnological purposes (see the entries on **meat, work,** and **experimentation** for further discussion of these issues). Other visible and powerful institutions in which anthropocentric attitudes and practices are dominant include such institutions as the **law** (wherein animals are denied status as full legal subjects) and education (in which entire branches of the university system such as the humanities and social sciences are grounded in ideologies of human exceptionalism, while other branches such as the physical sciences routinely subject animals to painful and invasive experimentation). Anthropocentric attitudes are also at work in subtle ways in practices ranging from the use of animals for **entertainment** (where animals are often seen in purely instrumental ways) to the construction of our cities and roadways (which often fragment animal habitats and endanger animal lives; see the entry on **roadkill** for further discussion).

See also: anthropomorphism; race

Further reading: Crist 2018; Steiner 2005

ANTHROPOLOGICAL MACHINE

A concept developed by Italian philosopher Giorgio Agamben, the anthropological machine names the deep conceptual and political structures of Western culture (Agamben 2004). For Agamben, the West has been concerned first and foremost with articulating a conception of community and politics based on what is properly and uniquely human in opposition to what is animal and nonhuman. This process of human-construction, or what Agamben calls *anthropogenesis*, is carried out primarily through conceptual and practical means and varies in different historical periods. Along these lines, Agamben distinguishes between two primary kinds of anthropological machines, namely, modern and pre-modern. In the modern machine, humanness is constructed by isolating and separating out nonhuman elements from what is properly human (for example, within this discursive apparatus a comatose human being is characterized as a mere animal body, not fully human, because it is lacking quintessential human capacities for speech and consciousness). By contrast, the pre-modern machine works in reverse, viewing certain kinds of human beings (such as slaves or barbarians) as having been essentially animal all along and hence not genuinely human. Agamben examines the functioning of the anthropological machine in order to highlight its deeply violent and divisive nature. Rather than accepting human beings as constituted by a complex mixture of human and animal elements that cannot be neatly separated or distinguished, the anthropological machine insists on defining human propriety through violent processes of inclusion and exclusion. It might seem that the solution to the violence of this machine is to establish, once and for all, what separates human beings from animals and to do so in a broad and generous way that does not exclude any human beings from the concept of humanity or human institutions. But Agamben argues that providing another version of the **human/animal distinction** is something we should avoid. Instead, he suggests we should acknowledge that human beings are thrown into the world in such a way that they carry with them an indistinguishable combination of humanity and animality and hence have no pre-established essence or identity to assume. Furthermore, he argues human beings have to learn to negotiate the passages between these domains and figure out who they are without any secure guideposts. To adopt this more complex and less secure way of being in the world is, for Agamben, how we can begin the process of stopping the anthropological machine (and its attendant violence) and enter into a new way of life. In this new way of life, Agamben suggests we will no longer be concerned to distinguish human from animal and, as a result, we can come to establish new relations with ourselves, animals, and the **more-than-human** world.

Within the context of animal studies, Agamben's concept has been useful for recognizing that the human/animal distinction not only concerns animals but simultaneously functions to create distinctions internal to human existence itself. In other words, the anthropological machine helps to make sense of the ways in which attempts to determine human nature have often been paradoxically accompanied by parallel efforts at sub-humanizing and de-humanizing certain portions of the human population (along the lines of **race**, sex, citizenship, and so on). One shortcoming of Agamben's analysis from the perspective of animal studies is that it fails to attend in sufficient detail to the way in which the anthropological machine inflicts violence on the animal and **more-than-human** worlds. Although Agamben expresses interest in more pacific relations with animals, his work to date has been largely silent concerning the specific sources and forms of violence toward animals and how they might be effectively challenged.

See also: biopolitics; dehumanization

Further reading: Agamben 2004; Agamben 2014; Gilebbi 2014; Mackenzie 2011; Oliver 2009

ANTHROPOMORPHISM

Anthropomorphism is the attribution of human form to animals and other nonhuman beings. The term appears frequently in scientific discussions of animal behavior and cognition, where it is typically raised as a critical charge against researchers who suggest that a particular animal behavior is best explained by ascribing a quintessentially human mental experience to animals. The charge of anthropomorphism is raised because, in scientific contexts, the governing paradigm is to describe only observable animal behavior and avoid positing explanations based on unobservable inner states. Thus, if a researcher sees two animals being aggressive toward one another, it would typically be considered anthropomorphic to describe that interaction as two animals "fighting" because they are "angry" with one another. Such anthropomorphic assertions, the critic would suggest, simply cannot be verified, as we do not have access to the inner lives of animals.

Caution against anthropomorphism is, in many instances, an entirely reasonable stance to adopt. After all, animals do many things that resemble certain human behaviors but that may serve very different ends (for example, a primate who appears to be "smiling" and "happy" to the untrained observer might in some circumstances actually be displaying an aggressive stance). Some critics of anthropomorphism have also maintained that human cognition is fundamentally different in nature from animal cognition, thereby ruling out in advance any meaningful explanatory overlap. This attitude toward animal cognition is grounded in what is sometimes called the principle of *cognitive parsimony*, which suggests that a given behavior should be explained if possible using the most basic cognitive capacities possible, and that appeals to higher-order cognitive capacities (which are common among human beings) must be accompanied by compelling evidence of the necessity for such appeals. In practice, cognitive parsimony has tended to lead to the assumption that nearly all animal behavior can be explained without recourse to human-like mental and emotional states.

Some animal researchers, many of whom work in the field of cognitive **ethology**, have suggested that anthropomorphism is not nearly as much of an issue as its critics maintain. Instead of remaining tied solely to the principle of cognitive parsimony, these researchers suggest that *evolutionary parsimony*—the principle that if two genetically similar species act in similar ways, their mental states are probably similar—gives good reasons for believing that anthropomorphism is sometimes justified, especially in studying species that have deep evolutionary similarities with human beings. From this perspective, anthropomorphism can be useful if it is used judiciously and critically to generate testable hypotheses and to provide reasonable explanations of animal behavior and cognition. Ethologist Gordon Burghardt, who coined the concept *critical anthropomorphism*, argues (along with his colleague Jesús Rivas) that anthropomorphism is ultimately a problem only if it goes "unacknowledged, unrecognized, or [when it is] used as the basis for accepting conclusions [about animal cognition] by circumventing the need to actually test them" (Rivas and Burghardt 2002: 10).

Primatologist Frans de Waal has pointed out an additional problem with mainstream critics of anthropomorphism, namely, that their presumption that anthropomorphism is intrinsically a category error can lead to what he labels *anthropodenial*. De Waal defines anthropodenial as the *"a priori* rejection of shared characteristics between humans and animals when in fact they may exist" (de Waal 1999: 258). For de Waal, assuming that anthropomorphism is *always* questionable constitutes a willful denial of the "human-like characteristics of animals, and the animal-like characteristics of ourselves" (de Waal 2006: 65). Such anthropodenial, de Waal argues, is indicative of a stubborn resistance to the findings of Darwin and **evolutionary theory**, which provide overwhelming evidence for continuities among human beings and many animals at physiological, behavioral, and cognitive levels. For de Waal and other ethologists who adopt this perspective, the closer an animal species is to the human species in terms of their evolutionary history, the more reasonable and logical it is to posit anthropomorphic explanations and hypotheses to understand that species' behavior.

See also: language; mind

Further reading: Crist 1999; Daston and Mitman 2005; de Waal 2016; Mitchell, Thompson, and Miles 1997

ART

Animals are deeply entwined with art in multiple ways. First, animals are commonly the principal subjects of artworks. Second, animal bodies and byproducts are frequently used to make artworks. Third, animals themselves are sometimes considered to be artists and produce their own artworks. These manifold relationships form a key locus of interest for the field of animal studies, as they not only pose complex ethical and **ontological** questions about the nature of animal existence but they also open up novel perspectives on the emotional and social significance of human–animal encounters.

Some of the most remarkable works of art involving animals are prehistoric cave paintings in places like Lascaux, Altamira, and Bhimbetka. Although scholars differ on the ultimate significance of animals in these cave paintings, nearly all viewers who see them are moved by their beauty. The animals are depicted in a variety of situations by artists who passionately admire and carefully study their subjects. One of the more confounding aspects of the cave paintings is that many of the animals who were so lovingly depicted were also hunted, killed, and eaten by the very peoples who made the paintings. This raises the paradox of how artists who clearly admired the animals they painted could simultaneously kill and eat them (Bataille 2005).

Similar paradoxes reside in the ways in which animals figure materially in the composition of artworks. The hair, skin, and byproducts of animals have long been used as materials for artwork, but animal bodies are now figuring directly in artworks in increasingly controversial ways. For example, artists have recently presented artworks featuring: dead animal bodies cut apart and preserved in formaldehyde (Damien Hirst); living animals who have been genetically altered so as to transform their appearance (Eduardo Kac); and the remains of animals who were deliberately killed by the artist and then adorned in various ways in order to be photographed (Nathalia Edenmont). Some of these works have met with considerable public outrage and calls for censorship from pro-animal critics. Such artists claim, however, that their art raises important questions about the nature of human–animal relations and, for this reason, should not be censored or held hostage to standard ethical concerns. (See Aloi 2012, Baker 2013, and Broglio 2011 for rich discussions of the complex issues raised by these and related artists.)

Conversely, a number of artists have also used living and dead animals in their artwork to make avowedly pro-animal statements.

Some artists such as Sue Coe (1995) primarily use drawn representations of suffering animals in order to illustrate the horrors of **CAFO**s and other industrialized forms of violence against animals, while other artists such as Jo-Anne McArthur (2013) use photographs of living animals and accompanying narratives to document and raise awareness of animal abuse. Recently, a number of **roadkill** artists have emerged who use their artworks both to record the often-overlooked deaths of animals on roadways and also as a means of providing roadkilled animals resting places that are more beautiful and meaningful (see Desmond 2016 for a helpful overview of the significance of roadkill art).

Whether animals themselves are artists is a hotly contested issue. That many animal species naturally engage in forms of preening and decoration that qualify as aesthetic is an uncontroversial claim. But can animals use standard art materials to create paintings, drawings, sculptures, and the like? There are numerous documented examples of elephants painting on canvases with brushes, gorillas and orangutans working with a variety of artistic media, and horses and pigs painting and drawing. The question that animal art skeptics raise in this regard, though, is whether the animals who engage in such acts do so of their own initiative or whether they do so solely as a consequence of human training. If it is the latter, it might seem that animal artists are simply making art by rote and are unable to create artworks in any meaningful sense. While such questions are certainly worth considering, animal studies scholar Vinciane Despret (2016) suggests that it is less important to focus on whether the animal as an individual is an artist and more important to consider the remarkable assemblages that are formed between human beings and animals in these forms of artistic creation. Following Despret, many human artists today are less interested in whether animals can be independent artists and instead focus on what forms of art can be done with and alongside animals.

Further reading: Aloi 2012; Baker 2001; Cronin 2018

BECOMING-ANIMAL

Becoming-animal is a term used by philosophers Gilles Deleuze and Félix Guattari to name surprising and profound points of overlap between human beings and animals. In order to understand this term's full significance, it is necessary to situate it in the broader context of their overall philosophical project. As philosophers, Deleuze and Guattari understand themselves to be creators of concepts that can be used to resist the dominant social order, an order they believe to be constituted by unhealthy desires and "abominable sufferings" (Deleuze and Guattari 1994). At the heart of this social order one finds what are sometimes called "subject-positions," or normative and institutional ideals that tell human beings what kinds of "subjects" (or selves) they should be and how to live their lives. Deleuze and Guattari refer to these kinds of rigid identities as *molar*; such categories misleadingly encourage us to think of ourselves and the world in terms of static, constant, and distinct kinds, groupings, and species (for example, human as opposed to animal, man as opposed to woman, citizen as opposed to immigrant, and so on). In contrast to molar approaches, Deleuze and Guattari champion *molecular* modes of thinking and living that are more complex and variable and that seek out relations and "becomings" between beings and groups presumed to be in opposition.

It is in view of this project of destabilizing the existing order and developing new ways of thinking and living that the concept of becoming-animal functions in their work. Becoming-animal is possible, for Deleuze and Guattari, because the differences between human beings and animals are not absolute; in other words, the differences that reside on this border are differences of degree rather than kind (Cull 2011). (In adopting this conception of differences, Deleuze and Guattari draw close to Darwin and **evolutionary theory**, although they challenge key elements of evolutionary thinking elsewhere in their work.) Becoming-animal does not, however, imply an erasure of all differences between human beings and animals. The concept is not intended to suggest, for example, that human beings can jump from one "molar" species (in our case, *Homo sapiens*) to another (say, a wild horse, *Equus ferus*) and actually *be* a different kind of animal. Further, Deleuze and Guattari strongly insist that with becoming-animal they do not have in mind human beings imitating animals or resembling them in simple, familiar ways. Becoming-animal refers instead to those events when human beings, thanks to being affected in a profound way by animals, become something other than the molar subjects they believe themselves to be.

In order for becoming-animal to constitute a genuine displacement of standard human-centered perspectives and modes of existence, only becomings with certain "kinds" of animals will do. Here it is not so much a matter of specific animal species being particularly ideal for becoming-animal; any animal or group of animals might serve as the catalyst for a radical alteration. What matters is the particular *manner* in which the animal is encountered. Deleuze and Guattari refer to three different conceptions of animals along these lines: (1) *Oedipal* animals that are treated as familiar members of the family (such as **pets**); (2) *State* animals that correspond to familiar archetypes and carry standard symbolic meanings (such as we find in certain religious traditions); and (3) *demonic* animals that roam about in liminal spaces between discernible species categories (Deleuze and Guattari 1987). It is this latter "kind" of animal with which becoming-animal actually occurs, for the first two ways of encountering animals only confirm our established human-centered categories and perspectives. In undergoing becomings with demonic animals (and, again, such animals might include anything from a house cat to a wolf pack, depending on the specific kind of encounter), we unexpectedly find ourselves situated in a common zone with animals that are normally understood to reside in wholly distinct realms, separate from human beings.

An example of becoming-animal can be located in the event of being unexpectedly exposed to terrible animal suffering. In such moments, one sees a singular animal body that no longer takes its accustomed form: writhing, animated flesh that slips outside of the normal categories we use to conceptualize animals. We might sense in such suffering a point of overlap between ourselves and the suffering animal, not in the sense that human beings and animals are similar subjects who have a right not to be subjected to suffering, but in the sense that we are both en-fleshed creatures, exposed to suffering and whatever else might come our way. In his book on 20th century artist Francis Bacon, Deleuze suggests that Bacon's flesh paintings help us to notice this point of overlap between human and animal (Deleuze 2003). Bacon's paintings portray human flesh as almost fluid, as coming off the bones, as dis-organizing what would normally be an organized, intact body. To acknowledge that one is en-fleshed in this manner is to accept that one's body belongs to a zone of impure and impersonal affects beyond one's control; it is to catch sight of the fact that we as human beings occupy a shared space (what Deleuze and Guattari call a "zone of proximity") with animals and other en-fleshed beings in ways we normally disavow. Such events of being displaced from one's familiar subject-position are quintessential instances of "becoming-animal." They serve to open up human individuals to other possible becomings, affects, and perspectives.

Some critics of Deleuze and Guattari have charged that there seems to be little upshot for animals themselves when human beings become-animal (Haraway 2008; Mullarkey 2013). From this perspective, becoming-animal might allow human beings to become-other, but such becomings and alterations do not seem to change the "abominable" situation of animals themselves. Deleuze and Guattari insist, though, that "we think and write for animals themselves. We become animal so that the animal also becomes something else" (1994: 109). It would seem then, that becoming-animal is motivated by a desire to create possibilities for animals to become something other than what the established order allows them to be. In becoming-animal, not only does one's human subjectivity undergo a radical alteration, but one also comes to feel a certain shame in regard to inhabiting and benefiting from a social and economic order that relegates animals to their en-fleshed state in the worst possible way. A more affirmative relation with animal life would thus entail challenging the established order so that other ways of living become possible for animals themselves. Although Deleuze and Guattari do not develop a full account of what they have in mind with animals becoming "something else," it would seem to include some manner of animals, humans, and other earthly beings co-existing and flourishing beyond the established order of **capitalism** and **anthropocentrism**.

David Abram (2010) is one of the more influential theorists who has used the concept of becoming-animal for his own work. He acknowledges the influence of Deleuze and Guattari's terminology for his own project, but he also seeks to grant the term new meanings. For Abram, becoming animal (he tends not to hyphenate the term) is most fundamentally about human beings affirming and living with and through their animality. This conception of animality includes the embodied, animal, and biological heritage of human beings as well as their materiality and "radical immanence" (a concept he also shares with Deleuze and Guattari, and which is used as a contrast with otherworldly transcendence). To affirm animality and immanence is, for Abram, to remain faithful to the earth with no desire or hope for leaving behind the earth-bound fate of embodied creatures. What one ultimately finds in affirming and experimenting with becoming animal, Abram suggests, is an animal world that is lively, creative, and filled with unexpected possibilities and forms of **agency**.

See also: human/animal distinction; more-than-human

Further reading: Beaulieu 2011; Gardner and MacCormack 2017; Lawlor 2008; Stark and Roffe 2015

BIOPOLITICS

Biopolitics is a concept that calls attention to the centrality of life (*bios*) in the political sphere. This entry first analyzes how the term biopolitics is generally used in contemporary theoretical discussions and then examines how it is employed in the context of animal studies in particular. Although "biopolitics" has been used in a variety of ways, nearly all work on this theme today derives from the seminal analysis by theorist Michel Foucault (1990, 2003). Foucault argues that the concept of biopolitics is essential for tracking a noticeable shift at the beginning of modernity away from a political paradigm based on sovereign rule over subjects toward a politics focused on the formation of individual bodies and populations. The former, sovereign form of politics operates in a primarily repressive manner and has its foundation in a supreme entity (for example, a king) having the power to kill its subjects. Sovereign power of this sort corresponds with the familiar image of an individual leader or political body that stands outside and above society and rules through fear, sheer force, and threats of death. Biopolitics, by contrast, refers to a form of political power that functions in less overtly repressive, and even productive, ways. Power that takes this form (which Foucault refers to as *biopower*) is more subtle and supple, shaping and controlling bodies and populations in order to direct their living energies toward the goals of the state and other social, economic, and administrative powers. (As an example, consider instances when a national health administration is concerned with a particular disease outbreak less because of the harms it might cause to individuals in relation to their private lives and loved ones and more because of the loss of work hours and national economic productivity and growth.)

One of the paradoxes of biopolitics is that individual subjects often come to *desire* this kind of control. As individuals become habituated to various kinds of biopolitical formations, they eventually start to identify positively with the norms, ideals, and methods used to administer their existence. In this way, biopolitics operates not so much from above (as the classical picture of sovereign power would have it) but from below, with the controlled individuals and populations ultimately governing and regulating themselves. While such modes of control might appear inherently and entirely pernicious, Foucault argues that biopolitics and biopower are not simply evil or problematic. Rather, his reflections on these concepts are meant to help us gain a critical grasp on important shifts in the circulation of power within modernity and a keener sense for how biopower persists and is transformed in our own age. This updated conception of power can also help us better understand extant forms of political resistance and better appreciate how biopower relies on our vital energies to extend its reach and sustain itself. From this more complex and ambiguous perspective, one can see that biopolitics has a double-edged character: (1) a dark side of control, aimed at reducing the lives of individuals and entire populations to the role of providing services to the state and other apparatuses of power; and (2) an opposing, more promising side in which power is seen to operate only insofar as it is able to makes use of bodies and life energies that offer unexpected resistance to control and governance.

Contemporary theorists of biopolitics have tended to emphasize one or the other side of biopower. Giorgio Agamben (1998, 2004), for example, offers a largely pessimistic account of biopolitics. Contra Foucault, he suggests that biopolitical violence and strife is much older than modernity and has been the operative logic of the Western political tradition since its inception in ancient Greece. In Agamben's sweeping analysis, the political sphere in the West was generated by a fundamental split between animality or bare life (zōē) and human political life (bios). This split, which is produced by what he calls the **anthropological machine**, passes within and through human beings, segmenting off human animality from what is taken to be true and genuine human subjectivity. The social order that emerges from this split between animality and humanity enables (1) the political capture of those who are deemed full human subjects (in the form of making them citizens who are subjected to governance), and (2) the simultaneous exclusion of de-humanized beings (whether human or animal) from that same sphere. The paradox here, Agamben argues, is that although this biopolitical order excludes dehumanized beings from its domain (such beings are all deemed to be non-citizens), it still includes them within the purview of its power and ability to wield violence (such beings can be killed with impunity). Given the deeply violent nature of this biopolitical social order, Agamben maintains that the goal should not be to expand the biopolitical order to make it more inclusive but instead to invent an altogether different form of life that no longer produces these kinds of violent schisms and exclusions.

Michael Hardt and Antonio Negri (2000) present a very different account of biopolitics, seeing in the biopolitical control of human life the very conditions for undoing that order. As Hardt and Negri point out, the established social order is entirely dependent on the lives and productive power of the human beings that constitute and nourish it. Hence, as contemporary globalization works its way into nearly every aspect of human and planetary existence (to form what Hardt and Negri refer to as "Empire"), it simultaneously and increasingly inter-links those human lives and productive energies across the globe (a grouping they call the "multitude"). If the multitude of workers and producers who sustain Empire come to recognize the ways in which they themselves hold up the established order, they might then be able to wrest back autonomy over their productive energies and redirect them in more affirmative ways.

Despite the unquestionably important insights into the nature of modern power these biopolitical thinkers offer, their accounts tend to be unthinkingly human-centered and inattentive to the manner in which animals are also caught up within and resist biopower (Chrulew 2012; Shukin 2009; C. Taylor 2013; Thierman 2010). Recently, however, a number of animal studies theorists have sought to correct this oversight and explore various ways in which a bio-political analysis might illuminate both the main drivers of violence toward animals and novel forms of animal resistance and **agency**. Much of this work on what could be termed "animal biopolitics" takes a distinctly negative and critical stance. It highlights how bio-political logic extends its malignant reign beyond the sphere of in-terhuman relations deep into the lives and deaths of animals. Thus, just as biopolitics marks a shift in modernity toward the management of human life toward the administrative ends of the state and economy, the same shift can be marked in regard to the lives of animals. No longer are animals merely hunted in the wild or farmed in small numbers in domesticated settings; today, animal life is managed in massive numbers in highly technical, scientific ways in order to extract maximum profit from their lives, flesh, and even their left-over body parts (Shukin 2009). This kind of biopolitical man-agement of animal life and death is found most clearly in slaughter-houses and **CAFO**s, but scholars have made the case that biopolitics is at work in everything from contemporary **zoos** to animal **experi-mentation** to the **pet** industry (see Wadiwel 2018 for an illuminating overview of this work). The ultimate point of the bulk of these analyses is to better understand animal–biopolitical logic so that we might identify its pitfalls and develop more respectful ways of life and forms of individual and collective governance.

Other theorists offer a more complicated engagement with biopo-litics, seeing its development within modernity as something that has negative effects on animals and human beings but also as potentially providing means for a more affirmative politics of life. The most influential work of this sort comes from Cary Wolfe (2013). As with other scholars working on animal biopolitics, Wolfe hopes to place consideration of both human and animal well-being squarely at the core of biopolitical thought. In situating both groups within what he refers to as a "biopolitical frame," Wolfe is able to demonstrate that biopolitics captures human beings and animals in sometimes overlapping, sometimes distinct ways. Consequently, he suggests there can be no single, unified way of developing resistance to biopolitical violence. To be effective, biopolitical resistance must be differentiated and strategically refined. Toward this end, Wolfe experiments with the idea of an affirmative biopolitics that understands the joint capture of human and animal life as a challenge and an opportunity to think through a more expansive ethics, one that heals some of the fractures that harmful biopower creates within and among human and animal life. Wolfe argues, though, that this expanded and affirmative ethics and politics of life cannot reasonably include *all* life forms within its scope. What is at issue in ethics and politics from this perspective, he suggests, is to reconsider relations with beings who are *responsive*, that is to say, beings who are characterized by some sort of minimal subjectivity. These are the kinds of living beings who have the cognitive and affective equipment to co-exist with others in meaningful worlds. Wolfe's affirmative biopolitics thus includes a great number of more-than-human animals within its orbit—indeed, far more than are included by some versions of **animal welfare and animal rights**.

Wolfe's point about the importance of accounting for the differing effects of biopolitics on human beings and animals can be extended to suggest that biopolitical analysis as such is by itself an insufficient framework for scholars and activists interested in animal studies. Developing this point, philosopher James Stanescu (2013) argues that even though biopolitical frameworks uncover important ways in which biopower is applied to and circulates among animals, it is equally crucial to attend to the unique history of the capture and control of animal populations. Simply borrowing a framework that was initially developed to explain the logic of intrahuman violence and applying it to animals will, Stanescu maintains, fail to capture some of the specific ways in which animal life and death have been and continue to be captured and managed.

Further reading: Chrulew 2012; Lemke 2011; Neo and Emel 2017; Wadiwel 2015

CAFO

CAFO is an acronym for Concentrated Animal Feeding Operation. These operations (along with smaller AFOs, or Animal Feeding Operations) produce the majority of the meat, dairy products, and eggs consumed by citizens in the United States and are rapidly increasing their presence across the globe. They are chiefly characterized, as their name implies, by "concentrated" methods of animal feeding and other agricultural practices. CAFOs raise extraordinarily large numbers of animals in relatively small areas, making use of a variety of controversial enclosure and confinement techniques.

CAFOs are a fairly recent invention. In the United States, until less than a century ago the vast majority of meat, eggs, and dairy products were produced by large numbers of farmers on small farms scattered across the country. With the rise of CAFOs, the numbers of farms and farmers shrank, while the number of animals raised for consumption rose dramatically. This transformation was achieved by industrializing animal agriculture in ways akin to other forms of industrialized factory production (which is why CAFOs are often referred to as *factory farms*). Animal flesh and byproducts are produced on CAFOs with an eye toward maximum profit, speed of delivery, and economic efficiency; these aims, in turn, lead to the shrinking of human labor involved in the raising and production process, increased automation at every level, and feeding techniques that rapidly accelerate animal growth. The number of animals raised for human consumption in CAFOs per year is staggering—nearly ten *billion* per year in the United States alone. To gain some sense of the scale of these farms, consider the qualifications for a single farm to be considered a "large CAFO" by the United States Environmental Protection Agency. Such a farm would have on its grounds at any given time a minimum of 700 dairy cattle, or 1,000 veal calves, or 2,500 swine, or 30,000 ducks, or 125,000 chickens (EPA 2007).

What explains this shift in animal agriculture, from large numbers of farmers and smaller farms to such enormous and technologically advanced feeding operations? CAFO advocates argue that concentrated farming techniques were developed in response to the need to feed a rapidly growing population. In addition, they maintain that CAFOs have created a better world for consumers and workers, making more food available at a lower cost while freeing increasing numbers of people from the difficulties of a traditional agricultural way of life (see Imhoff 2010 for further discussion of this position).

Whatever the merits of such arguments, both defenders and critics of CAFOs alike acknowledge that a hyper-industrialized approach to farming creates a number of ethical problems in relation to the environment, human beings, and the farmed animals themselves. In terms of environmental and human health issues, CAFOs pose serious threats, due in large part to the excessive waste that is created in farming large numbers of animals in concentrated settings. A single large CAFO can produce more than 1.5 million tons of waste annually, which exceeds the amount of waste produced by some large cities (Hribar 2010). Often, untreated waste is liquefied and then spread back onto surrounding agricultural areas in amounts the land cannot reabsorb. As this run-off waste is subsequently integrated into the surrounding environment by way of air, land, and water cycles, both human and nonhuman entities can suffer ill effects. Surrounding water systems can become polluted, affecting the living conditions of animals and the drinking supply of human communities; high levels of ammonia and other gases can cause air pollution, affecting wildlife and nearby neighborhoods; antibiotics used to control disease outbreaks among farm animals can pollute waterways and have negative effects on the antibiotic resistance of those who drink polluted water; and methane emissions contribute significantly to global greenhouse gas emissions and the problem of **climate change** (see Kraham 2017 and Spellman and Whiting 2007 for further discussion of the environmental effects of CAFOs).

As troubling as these problems might be for the ecology and nearby human communities, they pale in comparison to the harms suffered by the farmed animals themselves. In order to raise animals in such large numbers and small areas, animals have to be held in highly confined and crowded conditions. Dairy cows are often confined to indoor pens that leave them little ability to move or turn around. Veal calves are usually chained in narrow pens and left in dark conditions the majority of the day. Sows (female pigs used for reproductive purposes) often live in confined conditions year round and are placed in cramped birthing pens before giving birth. Chickens raised for meat (called "broiler" chickens) are typically raised indoors in large, crowded buildings, often with tens of thousands of other chickens. Chickens who lay eggs (called "battery" or "layer" chickens) are held in small, stacked cages along with several other chickens in each cage, in quarters so cramped that spreading their wings is rendered impossible (for a more detailed account of animal conditions on CAFOs, see Singer 2001a: 95–158).

Such conditions create considerable frustration, stress, and pain for the animals, which in turn can cause aggression, serious injury, and death. Scientists and farmers have developed a wide array of controversial techniques for limiting these forms of inter-animal aggression and violence. For example: to avoid injuries among chickens, debeaking is used; to prevent pigs from biting each other's tails out of frustration, tail docking is employed; and to avoid serious wounding among cattle, dehorning is practiced. Even if these procedures are carried out properly with anaesthesia and under medical supervision (which is not always the case), they cause evident pain and raise fundamental concerns about animal welfare.

In response to these and other problems with the treatment and welfare of CAFO animals, animal advocates have developed a number of strategies aimed at reforming and even eliminating CAFOs. For animal *welfare* advocates, making improvements in the way animals are treated is the primary aim. This approach has led to the introduction of legislation aimed at better living conditions for animals (bigger cages, more room to move and walk around, and so on) and improving slaughtering methods. Animal *rights* advocates, by contrast, believe that CAFO agricultural practices should be eliminated entirely—for not only do they cause extreme pain and suffering for animals, they also employ a form of mass killing that reduces animals to mere commodities to be sold and consumed. Toward the end of abolishing CAFOs, animal rights activists have sought incrementally to eliminate certain CAFO practices, such as caging battery hens and the production of veal. This incremental approach is often coupled with vegan advocacy and activism that exposes the horrific living conditions of animals on CAFOs, in the hope that reducing consumer demand will make the CAFO business model unprofitable.

Animal agriculture corporations have responded in turn with their own legislative and economic strategies. They have sponsored legislation aimed at criminalizing undercover investigations of their farms (so-called "ag-gag" laws, aimed at "gagging" activists); in addition, they have sought to restrict speech that criticizes their agricultural practices and that can possibly cause them economic harm (so-called "veggie libel" laws) (see Moses and Tomaselli 2017 and Lovitz 2010 for a fuller discussion of these struggles between pro-animal activists and corporations). Given the significant amount of animal suffering, the number of animal lives and deaths at stake, and the widespread environmental and public health effects that arise in relation to industrial agricultural methods, CAFOs will almost assuredly remain a central point of scholarly and activist interest in animal studies for the foreseeable future.

See also: abolitionism; activism; animal–industrial complex; animal welfare and animal rights; capitalism; death

CAPITALISM

Capitalism is a manner of organizing economic and social relations such that the accumulation of profit and economic growth are the central aims. Typically, capitalist societies and economies are characterized by the widespread existence of markets, which function to foster the production, exchange, and consumption of goods and services. While historians of capitalism trace its origins to economic and cultural developments in late medieval Europe, it emerges in its most recognizable form during industrialization in the 19th century. Capitalism has since undergone fundamental transformations as it has matured and expanded across the globe.

When considering the main drivers of the exploitation of animals, there can be little doubt that economic activities play a central role. The question for animal studies theorists and activists, though, is whether the economy—and capitalism in particular, which is currently the world's most dominant economic system—is *primarily* responsible for the current state of animals in society (McMullen 2015). To answer this question, it is necessary to consider some of the more general ways in which animals are deeply enmeshed in human social relations and economic activities. Sociologist Ted Benton (1993) provides a useful list of human–animal economic relations, including the use of animals: for labor; for food; for **entertainment**; as objects of study for increasing scientific knowledge; as sources of wealth creation; as tools for policing and controlling human beings who disrupt social and economic life; as **pets**; and as status symbols.

Although all of these relations predate the origins of capitalism, animal studies sociologists such as David Nibert (2002) argue that the emergence of capitalism and the dominance of the profit motive has *intensified* their violent and exploitative nature. Furthermore, with the emergence of near-total corporate control over the economy and social life in the 20th century, animals have come increasingly to be seen (through both ideological means and sheer exertion of power) as mere commodities and resources, deepening the sub-humanization of animals and exposing them to ever more extreme forms of economic exploitation. Nibert's primary example of this process of capitalist intensification is the meat industry and the production processes in **CAFO**s, where animals are ruthlessly exploited for their flesh and byproducts in numbers and ways unknown to previous generations. Such intensified exploitation of animals under contemporary capitalism, Nibert notes, goes hand in hand with the increased exploitation of poor and marginalized human beings, who are often forced to labor in animal-related industries doing dangerous and difficult work for low wages. He argues that the capitalist system threatens the human and more-than-human world with complete ecological and socio-economic collapse and ultimately benefits only a handful of extraordinarily wealthy members of the capitalist class in the short term.

Anthropologist Barbara Noske (1989) and historian Jason Hribal (2003, 2007) make the case that the increased use of animals for their labor is another marked aspect of the intensification of animal exploitation under capitalism. Animals do all kinds of labor and service **work** without any relevant compensation, from providing energy for agricultural processes to serving as sources of mobility and protection (Coulter 2016). In being inserted (often forcibly) into these roles as producers and laborers, animals suffer from various forms of alienation, including being alienated from the products they make to being physically removed from their fellow animal companions and natural habitats during the time in which they labor (Noske 1989). Moreover, when animals are no longer deemed suitable for work, they are often killed and consumed by an economic system that seeks to exploit every bit of profit from their bodies and lives.

These examples are just a sampling of the ways in which animal life has changed for the worse under capitalism. Undoubtedly, counterexamples of quality-of-life improvements could also be noted (for example, increased longevity for certain domestic animals, or improved veterinary care, and so on). Overall, though, such improvements are overshadowed by the widespread negative impacts of contemporary capitalism on animal life. This consequence returns us to the question posed earlier, namely, is capitalism primarily responsible for the violence toward animals we see in nearly all large-scale contemporary societies? Further, are there ways to ameliorate this kind of violence within capitalist economies? Or does achieving better conditions for animals require instituting alternative economic systems?

Mainstream animal ethicists have often taken the approach that the principle of supply and demand can largely eliminate violence toward animals within the current economy. The key intuition here is that if consumers refuse to purchase a certain product (say, meat or milk), then markets will respond to consumer demand and stop producing those products. Others influenced by this approach suggest that pro-animal legislation can be used to eliminate a particular market (say, the production of fur or animal-tested cosmetics) without having to upend capitalism entirely. Deepening this line of thinking, some economists and legal theorists argue that overturning the property status of animals would be sufficient for solving many forms of economic exploitation while still maintaining a market-based capitalist economy (see McMullen 2016 for a fuller overview of these arguments).

Theorists of a more radical orientation maintain that such reforms to capitalism are insufficient for addressing the problems with current human–animal economic relations (Best 2014; Nibert 2002; Torres 2007). Even if certain products might be eliminated through consumer demand, these theorists argue that such results would only partially address the wide array of economic abuses of animals and workers and would ultimately fail to ameliorate the deeper socio-economic inequalities and ecological degradation produced by capitalism. Thus, they call for the development of economic alternatives to capitalism that engender more respectful relations with animals. While no single alternative has been universally championed by anti-capitalist animal advocates, a number of possible political economies—from anarchism to communism—have been suggested (Clark 1999). Perhaps the most fully developed alternative in this discussion is offered by Nibert (2002), who argues for socialism as an alternative. For Nibert, socialist economies are attractive from a pro-animal perspective in that they tend to remove the sources of economic insecurity that lead people to abuse and exploit animals. It should be noted, though, that none of the standard socio-economic alternatives to capitalism make the well-being of animals a priority. Hence, there still remains the need for theorists and activists to articulate from the ground up post-capitalist political economies that are genuinely non-anthropocentric.

See also: animal–industrial complex; biopolitics

Further reading: McMullen 2016; Nibert 2017

CAPTIVITY

Captive animals are found in a large number of institutions discussed in this volume, from **zoos** and circuses, to **CAFO**s and other sorts of agricultural settings, to more seemingly benign institutions such as **shelters and sanctuaries**. Animals are also forcibly made to **work** and are subject to invasive **experimentation** under conditions of captivity. In all of these cases, the issue arises of whether keeping animals in such conditions can be given a compelling justification—for even a minimal sense of ethical respect for animals would seem to suggest that forcibly holding them captive violates their autonomy. What is more, given that animals often resist and escape from captivity (Hribal 2010), it would seem that many animals would prefer to maintain freedom of movement. While the bulk of animal studies scholars are quite critical of many of these practices of captivity, complex questions persist about whether captivity as such is always a harm to animals and whether it should be strictly eliminated (Cochrane 2014).

Animal philosopher Lori Gruen is one of the leading theorists concerning the question of animal captivity and, like most animal studies theorists and activists, is generally very critical of practices involving the capture and control of animals. Gruen suggests, though, that there are some instances when holding animals captive might be the most ethically preferable thing to do (such as in the case of placing vulnerable animals in sanctuaries when they are unable to survive in the wild). In order to determine when captivity is ethically acceptable, Gruen believes certain conditions need to be met. In particular, captivity should only be permitted under circumstances where an animal's dignity and autonomy are acknowledged and where the relation between captive and captor is based on mutual care and respect (Gruen 2014a). Gruen's position is by no means universally accepted in the field, however, and some activists and scholars maintain that captivity is inherently disrespectful and should be entirely avoided (Rivera 2014).

Recently, scholars who are interested in mass incarceration have begun to reflect on shared forms of captivity used to control both human beings—especially people of color—and animals. The most detailed work on this intersection to date has been undertaken by critical geographer Karen Morin (2018), who offers painstakingly detailed analyses of overlapping structures of oppression and incarceration among animals and human beings. In particular, Morin posits deep interconnections between: the deaths of animals in slaughterhouses and prisoners in execution chambers; the use of animals and prisoners as unwilling subjects in medical and scientific experiments; the forced labor of animals and prisoners; and the caging and surveillance practices characteristic of zoos and prisons (for further reflection on this last point of overlap, see also Struthers Montford 2016). Although Morin and other scholars who work on these topics acknowledge that dissimilarities exist between the captivity of animals and prisoners, they believe that exploring the intersections between these oppressions can be useful for forming points of solidarity among political movements for animal justice and struggles against mass incarceration.

One of the lingering challenges for the animal liberation movement in regard to the ethics of captivity stems from its heavy reliance on carceral measures to address the mistreatment of animals (see Beirne 2009 for a thorough discussion of animal abuse and criminology). As legal scholar Justin Marceau (2019) has observed, many mainstream **animal welfare and animal rights** organizations find themselves in a paradoxical position on the issues of liberation and captivity, fighting for the liberation of captive animals by seeking to hold their human abusers captive in penal facilities. Marceau argues that for pro-animal discourse and activism to be consistent with its highest ideals, it must pursue animal justice without such heavy reliance on carceral institutions.

See also: race

Further reading: Gruen 2014b; Thomas and Shields 2012

CARNOPHALLOGOCENTRISM

Carnophallogocentrism is a concept used by philosopher Jacques Derrida to explain the importance of meat-eating and masculinity in the formation of subjectivity in Western philosophy and culture. According to Derrida, the West has been dominated by a conception of the human being as a masculine, speaking, self-aware, self-identical subject. Such a person is able to maintain his identity and projects—his autonomy—regardless of what kinds of interruptions and challenges to his integrity might arise. Derrida maintains that this notion of a centered, autonomous human self is both misleading and ethically problematic, for it underestimates the deeply relational nature of all selfhood as well as the importance of responding to Others who might interrupt one's life.

Derrida's critique of the classical and dominant notion of human subjectivity places him in the company of a whole host of thinkers in the so-called "Continental" or modern European philosophical tradition, insofar as the bulk of this discourse takes a critical analysis of human selfhood as its starting point. In contrast to most of his peers, however, Derrida attends carefully to the manner in which animals and animality (or animal-ness) function in the construction of traditional notions of human nature. To illustrate, if we think more carefully about classical notions of human subjectivity of the sort just described (autonomous, self-aware, etc.), we notice straightaway that human beings are supposed to be the *exclusive* holders of such capacities. Only human beings can be genuine selves, subjects with autonomy, agents capable of undertaking self-directed projects—or so it is often argued. But if we were to question this standard definition of what it means to be human (as Derrida and most Continental philosophers do), we would simultaneously be calling into question the traditional **human/ animal distinction** that undergirds our ideas about human nature. Thus, we would also need to think more deeply and carefully about how we arrived at our understanding of human existence in relation to animals and how we might develop a less dogmatic conception of human and animal life.

It is in the context of such a project that Derrida develops the notion of carnophallogocentrism as a way of describing traditional forms of subjectivity. We can make better sense of how these forms of subjectivity are carnophallogocentric by breaking down the term into its component parts: logo-centrism, phallo-centrism, and carno-centrism. Derrida's earliest writings sought to expose the *logo*-centric prejudice of the Western philosophical tradition (logocentric here refers to the privileges and priorities granted by this tradition to the rational, self-aware, self-present, speaking subject). Later, when his attention turned to issues dealing more directly with sexuality and gender, Derrida explored the linkages between logo-centrism and *phallo*-centrism (phallo-centrism here denotes the quintessentially virile and masculine aspects of Western social institutions and conceptions of subjectivity), leading him to coin the neologism *phallogocentrism*. And, finally, in his writings on animals, Derrida suggests that *carno*- should be added to *phallogocentrism* in order to emphasize the idea that the dominant conception of human subjectivity should be understood not simply as a fully self-present, speaking, masculine subject but also as a *flesh-eating* subject.

Thus, for Derrida, if we are to continue the philosophical legacy of the critical investigation of the human subject, we are required to attend to three distinct but overlapping registers:

- self-presence (the *logos* of self-mastery, reason, speech, and transparent, unmediated access to one's inner mental life);
- masculinity (the manner in which virile and masculine/phallic ideals dominate and permeate the socio-cultural order);
- carnivorism (the requirement of the literal and symbolic consumption of animal flesh, a commitment to **anthropocentrism**, the hierarchical ranking of human subjects over nonhuman animals).

See also: absent referent, *animot*, anthropocentrism, feminism

Further reading: Derrida 2008; Derrida 2009; Derrida and Nancy 1991; Derrida and Roudinesco 2004; Still 2015; Turner 2013

CLIMATE CHANGE

Climate change denotes a range of phenomena that result from adding greenhouse gases (such as carbon dioxide and methane) to the atmosphere. Climate change includes what is sometimes referred to as "global warming" (the accelerating warming of the earth's surface, atmosphere, and oceans), but also encompasses a wide range of effects linked to global warming, including: rising sea levels; changes in the location and intensity of precipitation; increases in the frequency and magnitude of natural disasters; changes in agricultural productivity; the shifting in range of plant and animal species; and loss of biodiversity. The rate at which gases are being emitted into the atmosphere by human beings has been climbing rapidly and exponentially for decades, with atmospheric carbon dioxide concentration reaching 405 parts per million in 2017, a number that exceeds any measured level for some 800,000 years in ice core records (Blunden, Arndt, and Hartfield 2018). This rapid increase in the amount of carbon dioxide and other greenhouse gas emissions has led to a corresponding rise in global average land and ocean surface temperatures. From 1880 (the date when modern recordkeeping began) to the present, the global average temperature has risen 1.8 degrees Fahrenheit, with the last decade being the hottest decade on record (Union of Concerned Scientists 2016).

There are three primary ways in which climate change has been addressed within the context of animal studies: (1) the contribution that meat production makes to greenhouse gas emissions; (2) the effects that accelerating climate change have on animals themselves; and (3) the manner in which climate change transforms human–animal relations. The first issue has received the most discussion in the literature, as the production of animal flesh for human consumption is a significant factor in anthropogenic (human-caused) greenhouse gas emissions. The livestock industry creates emissions in a variety of ways, ranging from the use of fossil fuels in transporting animals and running farm equipment, to the release of methane by ruminant animals, to nitrous oxide emissions from tilled soils. Precisely how much livestock agriculture contributes to warming trends is a highly contentious and politicized issue. Scholars have offered estimates suggesting that livestock are responsible for anywhere from 10–50% of total global greenhouse gas emissions. The most commonly used number in the scientific literature is produced by the Food and Agriculture Organization of the United Nations, which puts the contribution of livestock production to total greenhouse gases at 14.5% (Gerber et al. 2013). (Beef and milk production take up the bulk of these emissions, with pig, poultry, and egg production also making a significant contribution.) Even this fairly conservative estimate makes it clear that climate change cannot be adequately addressed without making significant changes to meat production and consumption. Indeed, some animal ethicists believe that the substantial contribution of meat production to ongoing climate change provides a compelling reason to adopt a vegan diet; and the bulk of the scientific literature supports the claim that vegan diets tend to be far more benign than meat-and-dairy-based diets in terms of climate impacts (see Nordgren 2012, Springmann et al. 2016, Scarborough et al. 2014, and Weis 2013 for helpful research on these issues).

As climate change has accelerated over the past few decades, the effects of such change on animals themselves has become increasingly evident and measurable. At present, efforts to study these effects are underway in regard to a variety of species. No single story about the effects of climate change on wildlife and domesticated animals can be told, insofar as the conditions for decline and flourishing on a rapidly changing planet vary widely depending on the species and its particular location. In general, scientists argue that climate change tends to intensify the challenges faced by species that are already in precarious situations. Such animals are often forced to change their range due to shifts in weather, food availability, and habitat fragmentation. As species move to new habitats, they often encounter new predators, new diseases, and sometimes crowd out existing species. Other species that are unable to move and adapt rapidly enough face **extinction**. With global warming and the concomitant acidification of oceans, entire trophic chains are being affected, with less food and nutrients available for animals that often cannot find other sources of nutrition. Warming waters can also affect breeding rates among certain species, while loss of sea ice due to warming waters presents challenges for animals that use sea ice for cover and habitat. Although some animal species will flourish in a rapidly warming and changing climate, many more will face serious challenges and possible extinction (see IPCC 2014 for an extensive overview of these issues).

Changes in range of habitat, availability of food sources, and ecosystem composition associated with climate change are also bringing human beings and animals into unprecedented modes of contact. New species of wildlife are being crowded into urban areas, creating competition for food sources and the possibility of the introduction of zoonotic diseases. As animal habitats are further fragmented, more animals are crossing roadways in search of new habitats and food sources, making more of them susceptible to becoming **roadkill**. Conversely, as human beings themselves are increasingly dislocated due to climate change, they find themselves encountering species with which they are largely unfamiliar, both on land and in the waters. Examples of such climate-related changes for both animals and human beings could be multiplied indefinitely. What climate change (and a simultaneously increasing human population) seem to be producing, in short, is what philosopher Margret Grebowicz (2017) calls a "claustrophobic" world—that is, a world wherein animals and humans increasingly live side by side in unaccustomed and crowded conditions. Whether human beings and animals can learn to live together more respectfully and joyfully under such conditions, or whether increased crowding leads to more violence and loss of biodiversity will undoubtedly form one of the leading questions and challenges for animal studies theorists and activists in the coming decades.

Further reading: Archer 2012; Cudworth 2011; Schlottmann and Sebo 2018; Schneider and Root 2001

COLONIALISM

Colonialism refers to the unwanted occupation of territory that is already inhabited by others. It typically involves colonizers settling on occupied territory either temporarily or permanently and exploiting the original inhabitants and land for economic gain. In academic and political discourse, the term colonialism is used most often to refer to the wave of Western/European colonialism that began during the late 15th century and eventually expanded to include regions throughout the Americas, Africa, Australia, and New Zealand.

The vast majority of historical analyses of colonialism focus on the means used to impose colonial relations, and the effects these relations had, on Indigenous peoples. Recently, however, scholars from the fields of history, ethnic studies, and animal studies have begun to consider more carefully the role of animals within colonial encounters. For colonialism did not just interrupt and change the course of human cultures; it also had profound effects on native animals as well as the animals associated with colonizers. Historian Virginia Anderson (2004) argues, for example, that colonialism in the Americas cannot be properly understood without taking animal life into account. As Johnson observes, colonizers did more than bring their beliefs and bodies to the "New World"—they brought with them an entire way of life, grounded in an anthropocentric worldview and involving heavy reliance on domesticated animal agriculture. In establishing this alternative way of life on colonized lands, settlers interrupted Indigenous cultures and their existing relations with animals. In many cases, such disruptions had lasting implications for Indigenous foodways and led to the displacement and even decimation of native animal species.

Similar historical accounts of the importance of animals in the establishment of colonial regimes can be told for all of the major instances of modern colonialism. Colonial displacements of Indigenous ways of life and human–animal relations are not, however, simple one-time events relegated to the past. Even in situations where colonial regimes have been effectively removed through decolonization, the effects of colonialism do not simply disappear. Postcolonial theorists and activists who examine these issues have noted that one of the key persisting consequences of colonialism is the **dehumanization** and animalization of formerly colonized peoples (Huggan and Tiffin 2010; Deckha 2018). In the process of establishing colonial regimes, settlers often portrayed Indigenous peoples as being closer to animals and less-than-human, arguing that their traditional ways of life were uncivilized and remained fixed outside of history in an animal condition. This animalization of natives allowed both for (1) the killing of Indigenous inhabitants when extermination was deemed useful, and (2) appropriation of native land and labor in cases when that was the desired outcome. For formerly colonized peoples, rejecting their animalized and dehumanized status was and still remains an overarching priority (Fanon 1967a). Although many animal studies theorists share in the ethical and political aims of this liberation project, there are concerns that overcoming dehumanization by reinforcing human **ontological** and ethical priority over animals leaves animals themselves in a space where they remain vulnerable to colonial and anthropocentric exploitation and violence (Deckha 2018; Singh 2018).

In many places, though, colonialism has done more than leave a painful legacy—it continues in explicit but modified forms. In classical iterations of colonialism, colonists would invade and exploit foreign regions but ultimately retain and return to their home territory. But in some cases, settler colonists have permanently detached from their home territories and built new nations on occupied lands, displacing Indigenous peoples and claiming sovereign ownership of their land. Settler colonial regimes of this sort exist today in such nations as the United States, Canada, Australia, and New Zealand (for a helpful overview of settler colonialism, see Veracini 2010). In response to ongoing settler colonialism, there has been a resurgence among Indigenous peoples of calls for decolonization as well as for a reconsideration of how Western-European forms of life are involved in dehumanization and violence toward animals and the **more-than-human** world.

Contemporary advocates for decolonization within settler societies are at pains to insist, though, that decolonization is more than a metaphor for changing consciousness or making minor institutional reforms to the existing social order (Tuck and Yang 2012); in many cases, decolonization is understood to involve challenging the political and legal legitimacy of settler colonial nations and working for the repatriation of Indigenous lands. This radical vision of decolonization poses complex challenges to the field of animal studies and projects for animal liberation. As Billy-Ray Belcourt (Driftpile Cree) has observed, many of the aims and methods characteristic of pro-animal discourse and activism presuppose and normalize the legitimacy of settler nations (Belcourt 2014). For example, Sue Donaldson and Will Kymlicka's call for developing a **zoopolis** that integrates domesticated and wild animals into the extant political order effectively suggests that settler states can be reformed to accommodate the interests of animals while doing justice to the aims of Indigenous peoples. But for decolonization advocates like Belcourt, justice for both animals and Indigenous peoples—and for the land more broadly—cannot be achieved within the economic and political structures of modern settler nation-states. What is needed, he suggests, is the reconstitution of human–animal relations based on Indigenous ethical principles and ways of life that are often dramatically at odds with the settler colonial societies (see also Watts 2013 for further examination of this point). Conversely, Belcourt maintains that, although the ideal of maintaining respectful relations with animals can be found within many Indigenous traditions, the bulk of the discourse and activism surrounding decolonization has often failed to stress the importance of rebuilding human–animal relations in an ethical manner. Consequently, Belcourt (and other theorists and activists who adopt a similar stance) insists that concerns about animal flourishing must come to occupy a central position in modern decolonization struggles.

One of the most fruitful experiments in bringing together pro-animal discourse and Indigenous forms of life is found in the work of Margaret Robinson (Mi'kmaq) (2014). Although **vegetarianism and veganism** are often characterized as being non-Indigenous, white, bourgeois practices, Robinson argues that the ethical ideals animating this alternative way of eating actually indicate a desire to establish more respectful relations with animals—a desire that, she suggests, lies at the very heart of her own Mi'kmaq tradition and several other Indigenous traditions. Robinson acknowledges that most Indigenous cultures have traditionally engaged in **hunting** and other means of killing and consuming animals but argues that many of these same traditions are explicitly premised on trying to live as respectfully as possible with regard to animals. Thus, if more respectful ways of eating become available (such as vegetarianism and veganism), and killing animals for food is no longer strictly necessary, she suggests it is incumbent upon those who live according to such ideals to consider altering their eating practices (on this point, see also Womack 2013). Although such drastic changes to traditional foodways might seem inauthentic and a betrayal of Indigenous traditions, Robinson argues that Indigenous forms of life are always in flux, responding to changing environmental and cultural conditions, and revisiting the question of what it means to live well and be respectful of all relations. Indeed, for Robinson, such responsiveness and flexibility are the main reasons Indigenous traditions like hers have been so resilient in the face of colonialism. From this perspective, practices like veganism and vegetarianism might possibly function not just as respectful ways of relating to animals but also as key strategies for survival in a world of declining human health, rapid ecological degradation, and loss of biodiversity.

Further reading: Few and Tortorici 2013; Johnson and Larsen 2017; Parreñas 2018

COMPANION SPECIES

Companion species is a concept used by Donna Haraway to describe the complex, interwoven, and deeply relational nature of earthly life. Although most of the examples she uses to illustrate companion species are based on human–animal relations, Haraway emphasizes that the concept is meant to encompass relations among species and kinds of all sorts, both human and **more-than-human**.

Although the term companion species appears prominently in Haraway's 2003 book, *The Companion Species Manifesto*, it can perhaps best be understood as deepening and bringing to fruition central theoretical and ethico-political themes in her earlier and widely read essay, "A Manifesto for Cyborgs: Science, Technology, and Socialist Feminism in the 1980s" (Haraway 1985). In that work, Haraway sought to develop a notion of human subjectivity as constituted by numerous relations and influences that have traditionally been thought to reside in a domain distinct from human existence. In particular, she argues that the distinctions between human and animal, human and machine, and physical and nonphysical have now been thoroughly breached; consequently, human beings must be under-stood as being co-constituted with and by animals, machines, and a wide variety of physical and non-physical relations and forces. This messy, complex notion of human subjectivity makes it necessary to discard the traditional conception of the human individual as a distinct and cleanly separated organism and replace it with an image of the individual as a cyborg—that is, a hybrid creature that has been as-sembled in and through various human and more-than-human beings and influences.

In the writings that follow this essay, Haraway turns her attention increasingly to human–animal relations and to further dismantling the

human/animal distinction. For Haraway, human and animal species do not constitute two distinct sets of beings but rather form complex relations that undercut any sense of purity in either grouping. Thus, human beings and animals are better understood as *companion species*, beings who share existence and who co-constitute one another in innumerable ways. In *The Companion Species Manifesto* and her follow-up volume, *When Species Meet* (2008), Haraway develops a robust notion of companion species that enriches and complicates traditional understandings of both the terms *companion* and *species*.

In using the term *companion*, Haraway does not simply have in mind what are typically called "companion animals," that is, **pets** who are seen as accessories for human enjoyment or as sources of **entertainment**. Rather, she wants to articulate a sense of companionship that is more profound and more challenging. For Haraway, companionship is a fundamental **ontological** category. Below companionships there are not atomistic, isolated individuals; rather, individuals emerge out of and thanks to relationships and companionships of various sorts. Companionship, then, is the most fundamental unit of existence and is basic to earthly life and death. Our companions—whether human or more-than-human—consort and break bread with us (Haraway often plays on the etymology of companion, from the Latin *com-panis*, meaning one who breaks bread with another). Together, companion species build meaningful worlds, lives, and societies. To share life with others, however, also means to share death with them. Relationships and companionships are built among beings who are finite (beings who are born into a world not of their own making) and mortal (beings who will die and decompose). For example: my own life is sustained thanks to the lives and deaths of others; and when I die, I will decompose and be returned to the earth, eventually becoming reconstituted by and among other organic and inorganic beings and relations. To be a genuine companion, a true comrade, is to care about these life-and-death relations and to learn how better to inhabit them. For Haraway, one of our chief political tasks is to give deeper thought and attention to how our social practices and institutions frame certain kinds of beings and groups as having value while rendering others unimportant and "killable" (that is, subject to violence and death with impunity).

Although best known as a feminist theorist, Haraway is a trained biologist, and her use of the term *species* is meant to recall readers to the frameworks of biology and **evolutionary theory**. While common sense notions of species tend to suggest species are fixed, static, or natural kinds of some sort, Haraway insists that they should be understood as dynamic and deeply relational in nature. In her account, species are characterized as groupings that emerge under specific historical and evolutionary pressures and, as such, are contingent (that is, the groupings could have been otherwise had they evolved under different conditions and will of necessity become something else over time as relations and selective pressures change). So, a species is itself a "multispecies crowd," internally differentiated and externally related to numerous species at micro- and macro-levels. This vision of radical difference and relationality is essential to Haraway's notion of companion species and is intended to undercut not just the idea of species as fixed natural kinds but also the notion of human exceptionalism. The human species, too, is a multispecies crowd, co-constituted by numerous species and relations, from gut microbes to other organisms to large-scale ecosystems. As with the varied meanings of companion, Haraway emphasizes the rich etymological senses of the word species, which range from wealth and filth to beholding and looking. This latter register of terms having to do with the act of looking (from the Latin, *specere*) are especially important for Haraway, because they point toward the ethical dimension of companion species. To inhabit multispecies relations in a way that does justice to them and understands them as an instantiation of "the good" is to look thoughtfully and carefully at—that is, to hold in regard and re-spect (from the Latin, *re* + *specere*)—those relations.

Haraway's most extended reflections on companion species focus, somewhat controversially, on dog agility training. For Haraway, when done well dog agility training is not a simple act of a human being exerting power and dominion over a particular dog. Instead, it requires careful and thoughtful negotiation on the part of a human being with the heritage of a given dog species, and the development of a set of practices that foster play, joy, and creativity on the part of both dog and human trainer. And while Haraway insists that her notion of companion species has thoroughgoing ethical and political consequences, she does not believe that **vegetarianism and veganism** or eliminating animal **experimentation** on animals is necessary. Rather, she believes these kinds of practices have evolved over long historical periods and that we can re-inhabit them in ways that are more respectful of animal lives and deaths.

Haraway's notion of companion species and her related writings on animals and multi-species relations have engendered in equal parts affirmation and critical rejection. While several theorists have found rich possibilities for research and practice inside her work (Grebowicz and Merrick 2013; Kirksey 2014), others (especially theorists and activists influenced by critical animal studies and vegan ecofeminism) suggest that her work remains deeply speciesist and is inimical to the aims of animal liberation (Rossini 2006; Weisberg 2009; Jenkins 2012; Gaard 2017).

Further reading: Gane and Haraway 2006; Haraway 2016a; Nast 2006; Schneider 2005

DEATH

The deaths of animals figure prominently in a striking number of entries in this volume. Whether we are considering the billions of animals slaughtered annually for their **meat**, the hundreds of millions who are subject to **experimentation** or **hunting**, the millions who end up as **roadkill** or are euthanized in **shelters**, or the countless other animals who die in **captivity** or as collateral damage in **war**, it is evident that animal death is a ubiquitous factor in human–animal relations. Although some of these deaths are understood to have significance for human beings and for animals themselves, the vast majority are paid little mind by society at large.

In reflecting on the ethics and politics of human death and violence, philosopher Judith Butler (2009) makes an important distinction between lives that are considered grievable and those that are characterized as not mattering. For Butler, one of the ways in which violence tends to expand and go unchecked is when it is directed at populations whose lives and deaths are considered unworthy of grieving. While Butler's work is focused on how such violence happens primarily in the context of intrahuman war, her framework is equally relevant to the ways in which animals' deaths both do and do not matter within contemporary societies (Stanescu 2012; Donaldson 2015; Redmalm 2015). In line with Butler's logic, animal studies theorists and activists hope to ameliorate some of the lethal violence directed at animals by having us reflect more carefully on the nature and significance of animal death. Along these lines, a number of historians and archaeologists have shown that animal death (especially the death of **pets**/companion animals) has been of some concern across human cultures since ancient times, as is clear from burial rituals, **art**, and other artifacts that express love and care for dead animals (see DeMello 2016 for an excellent collection of articles on these themes). A related phenomenon in contemporary societies is the growth of pet cemeteries, which offer human beings the opportunity to bury and grieve for their companion animals in a ritualized manner (see Desmond 2016 for a helpful discussion of this trend). In addition, the emergence of the no-kill movement in shelters testifies to the beliefs among some animal advocates that the untimely deaths of even unwanted animals matter and should be prevented whenever possible (see the entry on **shelters and sanctuaries**).

These displays of love and concern for the deaths of animals tend to be limited, however, to companion and domesticated animals or those animals who have benefited human beings in some important ways. For the billions of other animals who are not considered worthy of grieving, their deaths often go unnoticed and unremarked. In response to this uneven recognition of the significance of animal death, scholars and activists have undertaken various approaches to encourage people to attend more carefully to overlooked animals. Examples of such efforts range from underground activists who risk long-term imprisonment in order to uncover the horrific conditions in **CAFO**s and slaughterhouses (Park 2006) to artists who use their work to commemorate the deaths of roadkilled animals and animals who are euthanized in shelters (Desmond 2016; McArthur 2016).

Attending more thoughtfully to the deaths of animals raises the question of whether animals themselves have a relation to death, including both their own deaths and the deaths of others. In other words, do nonhuman animals themselves understand what death is? In the dominant philosophical and intellectual tradition of the West, it has long been maintained that human beings are distinct from other animals in our exclusive awareness of death. Presumed to be utterly lacking in **mind** and **language**, animals have been characterized as being unable to conceptualize the transition from life to death and thereby as cut off from any understanding of death or being affected by it in a fundamental way. This traditional stance on animal death is present even in Peter Singer's early animal liberationist philosophy, in which he explicitly argues that only animal suffering matters ethically and that animal death is largely irrelevant insofar as animals do not understand what death is (Singer 2001a).

Over the past several decades, though, our understanding of animals' relation to death has been enriched by researchers in **ethology**, anthropology, and animal studies. Ethologists such as Jane Goodall

(1990) and Frans de Waal (2019) have argued that chimpanzees, in particular, have complex responses to the deaths of members of their troop. Recognition of death among animals, however, is far from limited to chimpanzees and other primates. Behaviors ranging from corpse-carrying to mourning rituals to evident depression and grieving in response to the death of kin have been documented in several animal species, providing persuasive evidence that many animals have a refined awareness of and complex relation to death (Bekoff 2007; Hutto 2014; King 2013; Moss 1988). Thus, even if animals do not have precisely the same sorts of relations to death that human beings do, it no longer seems tenable to suggest that awareness of death constitutes a strict dividing line between human and animal (Derrida 1995).

Several important thinkers related to animal studies—including Val Plumwood (2012), Cary Wolfe (2010), and Bernd Heinrich (2012)—have argued that, although many human beings believe they have exclusive knowledge of what death is, we often fail to accept and acknowledge our deaths in an authentic manner. They further suggest that this disavowal indicates a certain desire among human beings to remove themselves from a shared condition of mortality with animals, thereby encouraging human disregard for the significance of animals' deaths. Thus, even as we blithely acknowledge that nonhuman animals will die, decay, and decompose, we tend to do what we can to avoid suffering a similar fate ourselves, including delaying death through technological and medical means, embalming dead bodies, and being buried in lead-lined coffins. These thinkers argue that one of the key steps in developing healthier and more respectful relations with animals lies in acknowledging our own animality and affirming the life-and-death circumstances we share with our planetary kin.

See also: becoming-animal; biopolitics; human/animal distinction

Further reading: DeMello 2016; Donaldson and King 2019; Johnston and Probyn-Rapsey 2013; King 2013

DEHUMANIZATION

Dehumanization refers to a set of practices and a mode of thinking that treat certain persons as lacking distinctly human characteristics. On the surface, dehumanization would seem to have little to do with animals, but recent research in psychology suggests important linkages between lay conceptions of humanness on the one hand and animal existence and nonhuman beings more generally on the other. Nick Haslam and Steve Loughnan (2014) have posited that dehumanization proceeds by way of two chief sets of distinctions: the **human/animal distinction** and the human/object distinction. With regard to the latter distinction, commonsense conceptions of human nature tend to perceive full human beings as being sharply different in kind from nonhuman objects, such as robots or machines. Consequently, outgroups or individuals who are deemed not fully human are often associated with machine-like behaviors and characteristics (for example, being cold, indifferent, and lacking subjective awareness). In terms of the human/animal distinction, dehumanization proceeds by characterizing certain persons as essentially animal in nature, denying them rationality, intelligence, and selfhood. Haslam and Loughnan, suggest that this "animalistic" form of dehumanization is often at work in genocidal practices, where out-groups of human beings are exterminated on the basis of being "mere" animal pests or vermin.

While much of the research on dehumanization is focused on trying to understand the origins of intrahuman marginalization, it also has important implications for understanding human violence and discrimination directed at animals. These implications can be seen more clearly if dehumanization is understood as being something fundamentally different from simply denying persons biological species membership. As philosopher David Livingstone Smith (2013) argues, dehumanization is a matter of treating certain persons as *sub-human* or *less than human*, where human refers to "my kind" rather than to a biological species category. Similarly, in regard to animals, Smith suggests that justifications for violence, exclusion, and discrimination are based not so much on **speciesism** (that is, discrimination against animals due to their lacking human biological species membership) but on the notion that animals are not "my kind"—which is to say, animals are deemed sub-human or less than human in much the same way that marginalized human persons and groups are characterized. If this analysis is correct, it suggests that there are deep intersections and similarities between the mistreatment of animals and the marginalization of persons on the basis of such traits as **race**, sex, **disability**, sexual preference, and related categories, insofar as all of these groups are subjected to a similar "logic" of dehumanization. Recently, dehumanization researchers have begun to examine the ways in which violence and discrimination against animals might be ameliorated by "humanizing" our perception of them and demonstrating how animals are similar to human beings in various ways (Bastian et al. 2012); conversely, emphasizing human–animal similarity has also been shown to be potentially relevant for rectifying discrimination against various marginalized human groups (Costello and Hodson 2009).

See also: anthropocentrism; anthropological machine; intersectionality; queer

Further reading: Bain, Vaes, and Leyens 2014; Haslam 2006; Kasperbauer 2018; Mackenzie 2011

DISABILITY

Disability advocates tend to focus on the manner in which the dominant culture perpetuates an ideology of ableism (discrimination favoring able-bodied persons) that leads to the marginalization and exclusion of people with disabilities from important cultural institutions; pro-animal advocates, by contrast, offer a thoroughgoing critique of the ideologies of **speciesism** and **anthropocentrism** that give rise to myriad forms of violence directed at animals. But what might the ideologies and institutions of ableism have to do with speciesism and anthropocentrism, and vice versa? In recent years, a number of theorists and activists have sought to demonstrate that there is, in fact, fundamental overlap between the concerns of these two fields in terms of both what they critique and seek to resist. It has been further suggested that the disability and pro-animal advocacy movements would jointly benefit from a fuller dialogue.

Over the past decade, scholars and activists such as Sunaura Taylor, Mel Y. Chen, and Anthony Nocella have made extensive cases for the position that the oppression of animals and people with disabilities intersect along various lines. As Taylor observes, disabled persons and nonhuman animals are often oppressed by similar social, economic, and institutional forces (S. Taylor 2013). Insofar as the dominant culture is organized around the interests of so-called "normal," able-bodied human beings, then disabled people, animals, and other beings who are considered abnormal or who diverge from the normative ideal of humanity are potentially subject to varying degrees of violence and marginalization. In Chen's terms, there is an "animacy hierarchy" at work in the dominant culture, where maximal value is ascribed to fully animate, able-bodied, white males and lesser value is ascribed to human and nonhuman beings to the degree that they depart from this normative ideal (Chen 2012). In view of the manifold intersections among disabled individuals, animals, and the **more-than-human** world, Nocella has sought to establish a distinct field of research and activism on "eco-ability," which seeks to understand the joint forms of subhumanization and **dehumanization** to which these beings are subjected while also developing affirmative visions for more respectful co-existence (Nocella, Bentley, and Duncan 2012; Nocella, George, and Lupina 2019).

As this kind of intersectional work has developed, scholars and activists have made efforts to reform both disability studies and animal studies in view of newly gained insights. Thus, it has been pointed out that pro-animal authors often employ highly offensive language in arguing for better treatment of animals by, for example: comparing the cognitive abilities of certain animals to so-called "marginal case" human beings (a term often used in the literature to refer to human beings who are severely cognitively disabled); using terms like "freaks" to describe counter-cultural vegan practices; or referring disparagingly to people who reason inconsistently about animal rights as "schizophrenics." Conversely, theorists in disability studies have begun to attend more carefully to the manner in which uncritical appeals to the shared humanity of able-bodied and disabled persons can work in unintended ways to marginalize animals and other beings deemed nonhuman. Beyond these critical points, several authors who work at the intersection of animal and disability studies have suggested that both fields point affirmatively toward a shared, alternative vision of ethical life. Whereas traditional ethical frameworks are often grounded in some human capacity, power, or ability (for example, rationality or the virtues), many animal and disability studies theorists argue for the need to develop an **ethics** centered on often overlooked values and conditions, including vulnerability, interdependence, and compassion (Wolfe 2010; Taylor 2011).

See also: intersectionality

Further reading: Carlson 2018; Crary 2019; Taylor 2017

EMPATHY

Empathy is defined as the ability of an individual to imagine or feel what another individual is experiencing. The term usually involves reference to appreciating and having awareness of the *emotional* states of others and being able to take the *perspective* of others into account. The term is important in animal studies for two main reasons: (1) it has been suggested by some theorists that empathy should serve as the ground of ethical relations with other animals; and (2) scientific evidence suggests that some nonhuman animals display empathy both for members of their own species as well as members of other species.

In mainstream versions of animal **ethics** developed in the 1970s and 1980s, empathy played a limited role. Although animal ethicists usually acknowledged that animals had their own emotional and subjective lives, it was not thought to be ethically important for human beings to empathize with animals. Instead, the leading idea among ethicists was that *reason* required us to extend ethical treatment and **equal consideration** to animals—regardless of whether we cared for or felt empathy for animals—because human beings and animals were sufficiently similar in ethically salient ways. Animal ethicists and animal studies theorists inspired by **feminism** argued that this kind of identity-based and reason-based approach to ethics was insufficient to generate needed changes in our attitudes toward animals. Instead, they maintained that empathy for animals was the foundation upon which to transform our interactions with animals and cultivate more caring and loving relationships with them.

Recently, animal studies theorist and philosopher Lori Gruen (2015) has further developed this feminist and empathy-based approach to animal ethics with her concept of *entangled empathy*. For Gruen, entangled empathy is an ethical ideal that combines both cognition and emotion to arrive at an understanding of the other's situation and assist others with their problems. This sort of affective responsiveness to others in need presumes we are already in relation with others—both human and animal—and that the key question is how to maximize care within those relations. It also presumes a willingness on the part of the one who cares about others to learn as much as is possible about the others one impacts. This point is especially important when we are considering how to interact with and respond to animals whom we might not understand very well. In such cases, to act empathetically requires learning as much as possible about the **world** an animal inhabits, while also relying on ethologists and other experts who have more familiarity with a given species for clues about what respectful and caring relations might entail. Entangled empathy should also function, Gruen suggests, to help us reflect on and contest the blockages that limit empathy, such as the various structural and ideological factors that encourage us to treat animals as mere machines devoid of emotion and subjectivity. As should be clear, empathy of this sort is something that is, at least in part, deliberately acquired and cultivated; thus, animal studies theorists who place a premium on this sort of empathy argue that we need to undertake practices whereby we learn better to empathize with animals, whether through formal education or through relationships with companion animals/**pets** (for more on the latter theme, see Paul 2000).

The ability to empathize with others and consider another individual's perspective has long been assumed by many scientists to be a capacity unique to human beings. However, in the past few decades, substantial scientific and anecdotal evidence has emerged to support the notion that empathy is not uncommon among various animal species. Laboratory experiments and ethological observations in wild settings have highlighted empathic behaviors among such varied species as rats, dogs, lions, dolphins, and many primate species (Bekoff 2007; de Waal 2009; Lents 2016). So common and visible are empathic behaviors among primates in particular that biologist Nathan Lents suggests "we can now confidently conclude that empathy is a universal feature of primates" (2016: 76). In the face of such evidence, skeptics might suggest that what appear to be conscious acts of empathy among animals are simply unconscious reactions that support the reproductive survival of one's own kin or species. What is particularly remarkable about animal empathy, though, is that it often crosses species boundaries and sometimes has no obvious survival value. Ethologists have collected dozens of accounts of cross-species empathy between, for example, elephants and antelopes, dogs and rabbits, and primates and birds (for an overview of this research, see Bekoff 2007 and Lents 2016). As we come increasingly to acknowledge that empathy circulates in various regions of the animal world, a central pillar of the **human/animal distinction** is simultaneously called into question. Whereas the capacity for empathy has traditionally been used to distinguish human beings from animals, the evidence now suggests that human beings have actually inherited it from nonhuman animals who pre-exist us by millions of years.

See also: ethology; mind

ENTERTAINMENT

The use of animals for human entertainment has a long and checkered history. Ancient authors attest to the use of animals in a variety of contexts, from the exhibitions of Egypt and Greece to the bloody and deadly circuses of Rome. Today, animals are used for entertainment in numerous contexts, ranging from film, theater, and television to **zoos**, rodeos, and circuses to underground baiting and fighting rings (Wilson 2015). Given the widespread abuse of animals in many of these settings, some prominent pro-animal advocates and authors have called for the complete abolition of the use of animals in entertainment industries. By contrast, other pro-animal authors suggest that the use of animals for the purposes of entertainment must be addressed in a more refined way, taking into account the potential benefits of such practices as well as the cultural contexts in which they take place.

One way to get at the ethical issues at stake here is to examine some of the more objectionable uses of animals in entertainment, such as in rodeos, circuses, racing, and staged fighting. In these settings, animals often undergo painful training regimens in order to become top performers and are commonly injured (for example, in racing and rodeos) or killed (for example, in fighting) during their performances. Given that such practices do not meet any genuine human or animal need, pro-animal advocates and the broader public more generally tend to support the elimination of such violent forms of entertainment.

However, these more exploitative practices represent only a fraction of the ways animals are involved in human entertainment. Other practices, such as having animals participate in shows, exhibitions, and agility contests, are arguably less exploitative. In fact, theorists such as Vicki Hearne (1986) and Donna Haraway (2003) argue that certain kinds of performances involving highly trained animals are actually *good* for both human beings and the animals themselves. Consider, for example, horses and dogs that have been specifically bred to excel at agility sports and their human companions. Such animals often actively desire to engage in training and performances; and learning how to participate properly in those performances requires human companions to learn to think from the animal's perspective and be more responsive to the animal's needs. Thus, these kinds of high-level performances might be said to constitute instances of human and animal co-flourishing (Haraway 2003). In response to this defense of animal entertainment, critics argue that animal sports are the result of an extended history of treating animals in an instrumental manner, from the practices involved in breeding high-performing animals to the domineering training techniques required to engender precise performances on the part of animals. Given this troubling history of domination, critics argue even these kinds of performances and practices should be abolished.

As the abolitionist perspective on animal entertainment continues to gain more hold within animal studies, it faces two difficult questions. The first concerns that of racism and cultural relativism. Sometimes the kinds of performance under critical scrutiny (for example, fighting and baiting) are associated with minority groups or with non-dominant cultures. For instance, the public outrage over the dogfighting operation run by professional football player Michael Vick (who identifies as African-American), or attempts by activists to prohibit culturally specific forms of bullfighting and rodeos (such as are found in Spanish and rural Mexican traditions), are sometimes seen as forms of culturally insensitive imperialism (see Kim 2015 and Boisseron 2018 for thoughtful analyses of these issues). The second question that arises for the abolitionist perspective is in regard to the genuine joy that many people experience in watching animals perform. If this animal-inspired joy is beneficial to human beings and likewise helps to engender love and respect for animal life (as Preece and Chamberlain 1993 argue), it would seem a pressing matter to find ways to meet this need for joyful animal encounters in less violent and less exploitative ways.

See also: abolitionism; companion species; pets

Further reading: Herzog 2010; Wilson 2015

ENVIRONMENTALISM

Environmentalism comprises theoretical frameworks and sets of practices aimed at establishing more sustainable and ethical relationships with the natural environment. Although environmentalism and the pro-animal movement share a fundamental interest in the flourishing of the **more-than-human** world, there has been little overlap between these two movements until recently. In fact, during the academic development of both fields in the 1970s and 1980s, their relationship was often tense and even outright hostile. In the pro-animal literature and activism of this era, environmentalism was either summarily dismissed or disparaged as being fascist in nature, due to its emphasis on the well-being of larger ecological systems and wholes (Regan 1983). Conversely, many environmental ethicists accused the pro-animal movement of being anti-ecological and overly concerned with the welfare of individual domesticated animals at the expense of the health of ecosystems and the planet as a whole (Callicott 1980). There is an important kernel of truth in both of these criticisms. Environmentalists of this era often paid scant attention to the horrific conditions under which domesticated animals lived and died and tended to reinforce **speciesism** in their personal lives and public policy recommendations. The animal liberation and rights movements during this time were, by contrast, concentrated almost exclusively on the fate of domesticated animals and thus remained largely inattentive to the fate of **wildlife** and a whole host of pressing environmental issues, including **climate change**, biodiversity loss, water scarcity, pollution, and so on (for more on these debates, see Hargrove 1992).

Over the past three decades, though, the activism and theory associated with both movements have undergone shifts that render visible fundamental points of mutual concern. In regard to animals, the predominant focus on animal ethics during the late 20th century has been supplemented by the interdisciplinary research on animal studies canvassed in this volume. Environmental discourse and activism has seen a parallel development away from a narrow focus on ethics and policy to broader discussions informed by recent work across the humanities and social and physical sciences. What has come into sharper focus in this recent work are the ways in which pro-animal and environmental concerns dovetail around a critical engagement with traditional notions of human nature. In rethinking the **human/animal distinction** and the nature/culture distinction, both approaches have helped to highlight the ways in which traditional notions of the human are constituted through and against various "nonhuman" others, including animals, nature, and certain segments of humanity who have been deemed to be closer to the nonhuman world. Thus, it is now widely recognized that both the environmentalist and pro-animal movements have a joint stake in contesting **anthropocentrism** and in replacing hierarchical and exclusionary anthropocentric worldviews with richer accounts of the embeddedness of human beings in animal and ecological relations (Plumwood 2002; Rose, van Dooren, and Chrulew 2017; Crist 2019).

Another development that has brought animal and environmental concerns together is the rapid rise in ecological degradation over the past few decades (see the entry on the **Anthropocene** for a fuller discussion of this issue). Ecological degradation tends to have its most severe and most visible effects on animal life, particularly in the form of biodiversity loss and species **extinction**. Moreover, it is no longer possible to pursue justice for animals as if global ecological issues were a separate matter; after all, the aims of animal liberation mean very little if animals are freed from captivity and violence only to live in habitats that cannot sustain their existence (Best 2014). As ecological degradation continues to intensify, causing animal habitats to shift and animal populations to migrate, animal studies practitioners are also faced with ever-more-complex questions about the status of wildlife in urban settings and the rebuilding of cities and urban infrastructure that take the well-being of animals into account (see Lorimer 2015 for extended reflections on these themes).

Beyond these points of critical convergence, there are also a number of potential benefits for both movements that might follow from a deeper mutual engagement. At present, neither environmentalism nor pro-animal advocacy have been successful at making the sorts of political and structural changes that are necessary to achieve their long-term aims. In many cases, though, the long-term aims of both movements overlap in essential ways; thus, there are good reasons for contingent and strategic coalitions to be established in these specific areas (for example, in regard to **extinction** and **climate change**). Further, insofar as both movements aim at building relationships that are more responsive to and respectful of the more-than-human world, their distinct approaches can be helpful in pinpointing lingering anthropocentric dogmas and practices outside their respective domains. To be sure, some of the aims of environmentalism and animal studies/advocacy will remain incommensurable for the foreseeable future; but complete ideological alignment is not necessary for successful coalitions to be formed. Indeed, in some cases, the persisting differences between these movements might even be productive for developing novel responses to changing global realities in an age of radical ecological change.

See also: Anthropocene; climate change; companion species; invasive species; rewilding; wildlife; zoopolis

Further reading: Alaimo 2016; Oppermann and Iovino 2017

EQUAL CONSIDERATION

Equal consideration refers to an ethical principle commonly invoked in discussions in animal **ethics**. The principle of equal consideration states that equal moral weight should be given to interests that are relevantly similar, regardless of who has those interests. This principle can be illustrated using a hypothetical situation involving the interests of two fictional human beings, Adam and Barbara. Both Adam and Barbara have an interest in attending their local state university, and both have earned the necessary grade point average and have met the requisite testing standards to be admitted to the university. But a rogue admissions officer decides that, while Adam's application for study at the university should be approved, Barbara's application should be rejected without being reviewed because she is female (even though the university is open to people of all sexual identities). In this scenario, Barbara's interests have been overridden and she has been treated unequally on the basis of her sex. Such treatment is unethical, one might argue, because for no compelling reason (sexual difference is irrelevant to the issue at hand) Barbara's interest in attending university has not been given the same or equal consideration as Adam's interest. If two people have an interest that is relevantly similar, there must be some compelling justification, the reasoning goes, to ignore or override one person's interest in favor of another.

Many animal ethicists argue that this kind of reasoning about equality of consideration should be extended not just to all human beings but to many animals as well (for influential examples, see: DeGrazia 1996; Francione 2008; Regan 1983; Rowlands 2002; Singer 2001a). The central idea here is that, if human beings and animals have relevantly similar interests, then the interests of animals ought also to be taken into consideration when making decisions that affect their lives. This extension of the principle means that if we engage in an action that causes animals pain or frustrates their preferences (such as experimenting on them, or using them for **work** or **entertainment** against their will), and we would not be justified in subjecting our fellow human beings to such pain or frustration, then we are effectively discriminating against animals without persuasive reasons for doing so. Such unjustified discrimination is referred to by many animal ethicists as **speciesism**, and the principle of equal consideration is often used to highlight and challenge this ethical prejudice.

Further reading: DeGrazia 1996; Garner 2005

ETHICS

A significant number of issues in animal studies are ethical or normative in nature—which is to say, they raise questions about what obligations we have toward animals and what a good life in common with animals looks like. The specific sorts of ethical questions that animal studies theorists and philosophers examine include the following: Are animals deserving of **equal consideration**? Is **hunting** animals permissible? Can animal **experimentation** be justified? Should animals be used for **entertainment**? Are **CAFOs** ethical institutions? Are there persuasive reasons for keeping animals in **zoos**?

In order to answer such questions, theorists and philosophers usually employ what is called a "normative theory." Such theories are intended to discern what ethics is ultimately about and, in some cases, provide tools for navigating ethical dilemmas. In this entry, only normative theories are examined and the specific applied issues (hunting, experimentation, factory farms, and so on) are left to their own entries.

With some notable exceptions, classical philosophers tended to exclude animals from the community of those to whom human beings have ethical obligations. But in contemporary ethics, there has been a concerted effort to rethink this exclusion, and a number of normative frameworks have been employed to this end. The most influential ethical approaches along these lines are: utilitarianism, rights theory, virtue ethics, and care ethics.

Utilitarianism maintains that ethical decision-making should be guided by a single norm: bring about "the greatest good for the greatest number." This motto requires some clarification, though. What, we might ask, is meant by "greatest good" and "greatest number"? What is "good" for utilitarians varies from thinker to thinker, but most often good is equated with having pleasure or satisfying preferences. As for the "greatest number," the number at issue concerns all who are affected by one's actions. Thus, utilitarianism tells me I should choose those actions that bring about as much "good" as possible for all those who are affected by my actions. Now, this might (and often does) involve causing some harm to others as well. The point of a utilitarian ethic, however, is always to maximize the *aggregate* good of all individuals affected by my actions. What makes utilitarianism particularly attractive to many pro-animal ethicists is that animals are typically included in the scope of beings who can be affected by one's actions. In other words, animal happiness and preferences figure directly in one's calculations for bringing about the greatest good for the greatest number. Classical utilitarians like Jeremy Bentham and John Stuart Mill were clearly aware of these implications of utilitarianism for animals and note as much in their work. Contemporary utilitarian philosopher Peter Singer (2001a) has expanded this approach into a robust framework for **animal liberation**. For Singer, rigorous application of utilitarianism entails that we should eliminate most forms of painful animal experimentation and the consumption of meat and animal byproducts—the argument being that these practices fail to increase the aggregate good of all the human beings *and* animals affected by them.

Rights theory and utilitarianism are often contrasted for their differing views on the ethical status of individuals. For utilitarians, the preferences and happiness of individuals do matter but are ultimately subordinated to the aggregate good. This means that, in practice, utilitarianism often requires us to sacrifice individual welfare to the greater good, leading in some cases to gross mistreatment of individuals who have done nothing to deserve it. Rights theory is grounded on the notion that such sacrifices are ethically problematic and that ethics ought to protect the inherent value of individual persons. Such protection is effectively what a "right" is: the protection of some essential aspect of a person's well-being (life, autonomy, privacy, and so on) from being violated for the greater good or some other collective interest. Associated with classical philosophers like Immanuel Kant and contemporary philosophers such as John Rawls, rights theory has had an enormous influence in ethical theory and in other domains of moral reflection. This view, at first glance, would seem to hold little promise for protecting the interests of animals, as rights are often thought to belong solely to *persons*—that is, conscious subjects who have **agency** and a stake in how their lives and deaths unfold. It has long been thought that only human beings (and perhaps not even all human beings) are persons, hence the justifiable exclusion of animals from the scope of those individuals considered to have rights. But for animal ethicists like Tom Regan (1983), recent scientific accounts of the subjective and cognitive lives of animals suggest that at least some animals *are* persons and hence are deserving of having their inherent value as individuals protected. For Regan, this notion of animal rights leads not to the *reduction* of the use of such animals for their meat or as experimental subjects or for entertainment purposes, but the total **abolition** of such practices.

Virtue ethics, which stems from a tradition dating back to Aristotle and the ancient Greeks, has become increasingly influential in contemporary philosophical ethics. Virtue ethicists assume that ethical life is sufficiently complex so as to undercut the possibility of coming up with any general guidelines or decision-procedures to help us navigate it. What is most needed in negotiating the complexities of ethical life, they argue, is a good character and a full understanding of what constitutes a worthwhile life. Virtue ethicists differ on what precise form the good life takes, but they generally agree that a worthwhile life must serve as the *telos*, or end, of our actions. In order to build and sustain a worthwhile life, it is not sufficient simply to abide by a set of ethical precepts; rather, one must come to embody a set of positive character dispositions, or virtues, that allows one to live well continually and persistently. Virtue ethicists have largely focused on the character formation of *human* beings and devoted scant attention to human–animal interactions and to animal flourishing. But some contemporary virtue ethicists such as Martha Nussbaum (2006) and Rosalind Hursthouse (2000) have made the case that it is entirely appropriate to ask how a genuinely virtuous person would treat animals; and they suggest that a fairly robust pro-animal ethic can be developed on these grounds, inasmuch as those of us who seek to flourish as human beings should take delight in the co-flourishing of animal life as well.

Like virtue ethics, *care ethics* is suspicious of the attempt to turn ethical theory into a calculating game or a search for a rational decision-making procedure that will solve every ethical conundrum. The primary aim of care ethics is instead to have us attend to the specificity and complexity of the ethical situations in which we find ourselves; and this means, in short, attending to the ethical *relationships* in which we are enmeshed. As such, care ethicists place **empathy**, compassion, and dialogue at the heart of their ethical framework and suggest that unless ethics is motivated by genuine concern for other individuals, unethical consequences will follow. Traditionally associated with women, affective concern for others has often been given less prominence in classical ethical theory (there are important exceptions here, such as David Hume); but theorists such as Carol Gilligan (1982) and Nel Noddings (1984) have argued persuasively for the importance of placing caring relations at the very heart of ethical life and that there are feminist implications in doing so. Although Noddings has argued that care ethics has a minimal role to play in our ethical relations with animals, Carol Adams (2015), Josephine Donovan (1990; 2006), and a number of other feminist theorists have constructed rich pro-animal frameworks from out of this tradition. The key thesis of feminist animal care theory is that ethical relations with animals must proceed from the perspective of maintaining and establishing caring relationships with animals, and that justice for animals emerges from such relations rather than being opposed to them (for further elaboration of these themes, see the entry on **feminism**).

Further reading: Armstrong and Botzler 2008; Taylor 2009

ETHOLOGY

Ethology can be generically defined as the scientific study of animal behavior. What differentiates ethology (at least in its classical form) from other scientific approaches is its emphasis on studying animal behavior as it occurs in its natural setting and in view of a given behavior's function and evolutionary history. Ethology thus attempts to understand animal behavior in a much broader manner than competing schools like behaviorism. For scientists influenced by behaviorism and similar reductionist paradigms, animal behavior is ideally studied in laboratory settings and with little regard to a given animal's ecology or evolutionary history. For ethologists, by contrast, attending to the ecological, evolutionary, and relational aspects of an animal's behavior is essential to arriving at a full understanding of its significance.

Contemporary ethologists trace the scientific origins of their field back to Charles Darwin, whose *The Expression of the Emotions in Man and Animals* (1872) explains and compares human and animal behavior in a proto-ethological manner. But ethology as a distinct field of academic inquiry only fully emerges in the 20th century, with the pioneering research of Konrad Lorenz and Niko Tinbergen (see Burkhardt 2005 for a helpful history of these founding figures). Tinbergen's "four questions" about animal behavior are still frequently used to guide ethological research, and they comprise the following themes: (1) What are the stimuli that cause a behavior? (2) How does a behavior enhance reproductive fitness? (3) How is a behavior modified over an individual's lifespan? (4) What is the evolutionary history behind a behavior? In adopting this multi-pronged approach, Tinbergen's framework suggests that multiple levels of scientific analysis should be integrated in order to arrive at a fuller understanding of a given animal's behavior (Tinbergen 1963). (See Bateson and Laland 2013 for an insightful contemporary review of Tinbergen's framework as well as an illuminating example of how this ethological approach can be used to understand the function, evolution, utility, and physiological mechanisms underlying bird song.)

Building on the work of these early ethologists, more recent practitioners have extended research on animal behavior into new registers. Gordon Burghardt (1997), for example, has called for adding a fifth question to Tinbergen's four questions, with the additional question focusing on an animal's subjective, private experience. The aim of such work, which is sometimes referred to as *cognitive ethology*, is to understand animal cognition and subjective experience "from the inside." In so doing, cognitive ethologists are taking up the challenge of answering the classical philosophical question posed by Thomas Nagel (1974) of "What is it like to be a bat?" or what is the subjective life of an animal like? For cognitive ethologists of this sort, animals are best understood not simply as biological organisms to be studied using third-person, objective perspectives and methods but also as agents in their own right with their own centers of consciousness and meaningful worlds. (For more on cognitive ethology, see the entry on **mind**.)

Unsurprisingly, scientists who study animal behavior with the subjective lives of animals firmly in mind often come to empathize with those animals and begin to see them as individual subjects who should be cared for and respected. Cognitive ethologist Marc Bekoff argues that the science of ethology should unabashedly affirm this kind of compassion for animals and, in the process, become a kind of "deep" ethology (Bekoff 1998). Bekoff uses the term deep ethology in analogy with deep ecology, which argues that ecology should involve more than a dispassionate scientific study of the natural environment and base itself on a profound love of and respect for the natural world. Bekoff suggests the same kind of ethical and emotive relation should ground ethology, thereby transforming the field into a loving, caring, and compassionate—but still scientific—investigation of animal behavior. Deep ethology of this sort would, he maintains, help both specialists and non-specialists alike to come to a fuller appreciation of the effects that human beings have on animals and make both groups more acutely aware of their ethical obligations and relationships with other animals. Influential examples of this kind of deep ethological practice can be found in the research of Jane Goodall (1986), Barbara Smuts (2001), and Joe Hutto (2014), among others.

See also: empathy; evolutionary theory; mind; world

Further reading: Crist 1999; de Waal 2016; Jensen 2002

EVOLUTIONARY THEORY

Evolutionary theory offers an explanation of how organisms evolve and adapt to their natural environments, as well as an account of how environmental and other pressures produce a wide diversity of species. Within the context of animal studies, evolutionary theory and its founding figure, Charles Darwin (1809–1882), are key references for thinkers and theoretical approaches that emphasize the biological continuity of human beings and animals. In a related vein, evolutionary theory is also important for developing naturalistic (rather than other-worldly or purely rationalist) ethical frameworks and for demonstrating that ethical sentiments and behaviors are shared among human beings and animals.

The dominant intellectual framework in Western culture for understanding the **ontology** of human beings and animals prior to Darwin has often been characterized as a "Great Chain of Being" (Lovejoy 1936), a framework that has its origins in ancient Greek philosophical and medieval and Renaissance Christian traditions. The Great Chain of Being places God at the top of the hierarchy of existence, as the most perfect and supreme form of existence, and then gradually progresses downward through other divine beings, different groups of human beings, and then to various animals, plants, stones, and minerals. With regard to the division between human beings and animals in particular, the break between these two groups was generally thought to be unbridgeable, inasmuch as human beings were produced by a special act of creation and thus carried certain divine attributes that other earthly beings lacked. The differences between animals and humans were considered inherent to the very grain of the universe, as having an effectively permanent status, and as decreed by divine order.

Over time, however, careful observation and documentation of the continuities and deep historical relationships between human beings and various animal species came to undermine this picture. Although the idea of natural evolution was posited prior to Darwin, it was his scientific research that provided a viable explanation of how evolution actually unfolds as well as an account of the basic processes that produced organisms and species. Darwin's work made the case that the evolution of species occurs through a process of natural selection, wherein offspring that naturally vary from generation to generation are "selected" on the basis of whether they are able successfully to reproduce in the context of certain environmental pressures.

Adopting an evolutionary framework radically undercuts claims to *human exceptionalism*. On Darwin's approach, there is no Great Chain of Being, no hierarchically ranked and ontologically separated classes of human beings and animals. Rather, there is one large, continuous but varying set of earthly beings who emerge from the same process of evolution through natural selection. In regard to the nature of human beings, evolution undercuts the idea that a special account of human creation is necessary. Human beings evolved through "common descent" with other animals, which is to say, we and other species descend from a common ancestor and have a shared evolutionary heritage.

Evolutionary theory also has profound implications for how we think about *differences* between human beings and animals and the *ethical obligations* we might have toward animals. In regard to the first issue of human/animal differences, the evolutionary picture suggests that any noticeable differences between human beings and animals are not fundamental, essential, or static. They do not constitute a difference in kind between humans and animals, only a difference in degree. As such, there are no absolute, insuperable ruptures or breaks between human beings and animals at any level, whether cognitive, behavioral, or physiological; and many characteristics and capacities that we believe to be uniquely human can be found to greater or lesser degrees among animals. Rather than being situated in a Great Chain of Being, then, human beings and animals belong to the same multi-branched tree of life.

In terms of thinking about ethical obligations between human beings and animals, evolutionary theory provides us with persuasive reasons for thinking that the project of restricting ethics to our fellow human beings based on some kind of shared, essential humanity is untenable. Many of the traits and capacities we consider deserving of respect in our fellow human beings (for example, autonomy or **sentience**) are found in varying degrees in a wide variety of animal species; the existence of such similarities raises the question of whether we should extend **equal consideration** to animals who exhibit these similarities. Further, evolutionary theory suggests that the main building blocks of ethics (such as **empathy** and certain pro-social instincts and forms of communication) can be found in several animal species. This picture suggests that ethics is not something that marks human beings off from the animal kingdom but is instead part of a shared human and animal heritage.

Some animal studies theorists have raised concerns about this vision of evolutionary continuity, arguing that it is reductive of human–animal differences (Derrida 2008). To examine human and animal differences and similarities exclusively through this kind of biological lens is, it is argued, to risk engaging in a kind of "biologism" or "continuism,"—which is to say, to risk reducing all kinds of knowledge to a strictly biological framework when other perspectives are equally important and insightful.

Further reading: Crist 1999; Darwin 1964; Dawkins 1993; Rachels 1991; Rodd 1990

EXPERIMENTATION

Animal experimentation refers to "any use of an animal for experimental or other scientific purposes which may cause it pain, suffering, distress, or lasting harm" (Council Directive 86/609/EEC 1986). Experimenting on animals is among the most important and contentious issues in animal studies and has generated an enormous debate in both the academic literature and public sphere. The issue is of direct relevance to nearly all of humanity today, as a substantial amount of the medicines, cosmetics, and potentially harmful substances that human beings use have been tested on animals. Using animals for such testing raises a number of serious ethical questions about the relative value of human beings and animals. Further, with rapid advances in biotechnology research, animals are being experimented on and genetically transformed in novel ways that undermine traditional normative paradigms.

Determining the precise number of animals used for medical, cosmetic, biotechnological, and related forms of invasive experimentation is a difficult task for two reasons. First, careful numbers regarding animal test subjects are not regularly maintained at either national or global levels. Second, even in the limited instances where fairly consistent data are available at national or regional levels, such data are often compromised by the fact that many animal subjects are simply not included in the counts. For instance, in the United States, the Animal Welfare Act does not require that animals excluded from the protections of the Act be included in reporting about experiments; what is more, the vast majority of animals used for experimental purposes (such as rats, mice, and others) belong to the category of unprotected species. With these problems in mind, the best estimates of the total number of animals used for experimental purposes at a global level are well in excess of 100 million per year (see Knight 2011 for a fuller discussion of the difficulties in arriving at such estimates).

Animals are subjected to a wide array of modes of testing and experimentation. They are used: to investigate how diseases develop and can be cured; to study addiction and other problem behaviors; to test drugs and consumer products for safety and toxicity; and for dissection and other educational purposes (see Monamy 2017 and Greek and Shanks 2009 for a detailed discussion of the full range of animal testing). While some forms of animal testing are minimally invasive and cause negligible pain, others are deeply invasive and cause long-term injury, persistent suffering, and even death. Some of the most notorious tests include the LD50 test, which establishes the acute toxicity of a given substance by experimentally determining a dose that is sufficient to kill 50% of the dose animal group. Another is the Draize test, which measures the irritancy of a substance or product by dropping it in a test animal's eye or applying it on exposed skin and examining the resulting effect (which could include anything from inflammation to blindness). Perhaps the most contentious kind of animal experimentation involves the use of primates who, because of their deep genetic and psychological similarities with human beings, have been used to study everything from maternal deprivation to substance addiction to autoimmune and degenerative diseases.

When challenged to justify animal experimentation in light of the severe suffering and extensive death it causes, experimental scientists and their advocates have marshalled a number of arguments in their defense. The primary case for continued animal experimentation is that it is effective and that countless medicines and life-saving/improvement technologies have been developed with animal testing. Further, it is argued that even though animals are biologically similar enough to human beings to make testing animals effective, animals remain sufficiently morally different from human beings to justify their use. Finally, advocates for animal testing emphasize that animal experimentation is carefully regulated according to the Animal Welfare Act and that proposals to undertake animal research at most facilities have to be approved by an Institutional Animal Care and Use Committee (IACUC) or similar review boards. Critics of animal testing contest each of these points, arguing that the benefits of animal testing are overstated, that animals are sufficiently different from humans biologically to render many experiments questionable and morally similar enough to human beings to render experimentation problematic, and that existing regulations on experimentation fail to take the bulk of experimental animals into consideration. (For a superb set of articles that covers opposing viewpoints on animal experimentation, see Garrett 2012.)

Some scholars have suggested that one way around the ideological divide between advocates and critics of animal experimentation is to apply the principle of the "Three Rs" to all future research that might involve animals. The goal with this approach is to: *R*eplace animals as experimental subjects with non-animal substitutes; *R*educe the use of animals where their elimination is not possible; and *R*efine experiments to minimize pain and death as much as possible (see Hubrecht 2014 for an extensive discussion of the Three Rs). Although the pragmatic approach is laudable in many respects, many **animal rights** and animal studies scholars believe that it fails to identify and challenge the deeper forces involved in turning animals into experimental subjects. What is needed, these scholars argue, is critical awareness of the historical, cultural, and institutional settings in which animal experimentation is rendered acceptable. Along these lines, animal studies scholars have called attention to the ways in which animal experimentation unfolds in socially constructed settings that profoundly shape the meanings we give to animal lives and deaths (Birke, Arluke, and Michael 2007), settings that encourage researchers to underestimate the sophistication of animal minds and overlook the value-laden nature of scientific research (Rollin 2017). Other scholars influenced by critical animal studies have examined the commodification of lab animals and the appropriation of their "labor" under contemporary **capitalism** (Nibert 2002; Clark 2014). Related to this research, scholars have also explored how animal experimentation reinforces a problematic **biopolitics** aimed at controlling both human and animal populations (Kirk 2017; Piehl 2017). While most of this literature leans in the direction of abolishing or at least radically limiting animal experimentation, some animal studies theorists believe that experiments with animals can be reformed in view of these concerns to be more respectful of animal lives and agency (Haraway 2008).

See also: ethics; law; mind

Further reading: Greenhough and Roe 2018; Garrett 2012; Knight 2011; Monamy 2017; Orlans 1993

EXTINCTION

The term extinction is usually employed in scientific discussions of animals to refer to the disappearance of a particular species. A species is listed as extinct when no members of that species are remaining or when that species no longer exists in its native habitat (as, for example, when only a few individual members of the species remain alive and are kept in **captivity**). From a biological perspective, speciation (the emergence of a new species) and extinction are normal aspects of evolution, with speciation typically outstripping the pace of extinction (hence, the earth's characteristic biodiversity). Extinction of individual species tends to happen at a very slow rate under typical evolutionary circumstances. This "background extinction rate" is difficult to determine with any precision and must be inferred from fossil records and other forms of indirect evidence (De Vos et al. 2015). What is clear from these records, though, is that the earth has seen previous periods where the extinction rate has spiked sharply resulting in mass extinctions (which are typically defined as the earth losing 75% of its species). There have been five previous mass extinction events, and many scientists suggest that the earth is currently in the middle of a sixth mass extinction (Barnosky et al. 2011), with the contemporary extinction rate being anywhere from 100 to 1,000 times the background rate (see Ceballos et al. 2015 for discussion of a range of estimates). What makes the present extinction rate especially noteworthy is that it is driven primarily by anthropogenic (human-caused) forces. The chief anthropogenic drivers of the current extinction of animal species in particular include: habitat destruction; pollution; overkilling and overharvesting; and **climate change**. All of these drivers are accelerating in terms of ecological impact, thereby threatening even more extinction of animal species and loss of biodiversity in the coming decades and centuries.

Although this scientific framing of extinction in terms of species and biodiversity loss helps to illuminate the scale and scope of the present crisis, some animal studies scholars argue that it glosses over the more profound meanings and consequences of extinction. As Eileen Crist notes, one of the great paradoxes of the present crisis is that it is occurring at the very same time at which we are beginning to gain a fuller appreciation for the richness and complexity of earthly life and animal **mind**s (Crist 2013a). Thanks to this increased awareness, many people are coming to recognize that a species going extinct is not simply the loss of an abstract concept but instead marks the end of an entire **world**, the annihilation of an intricate and irreplaceable way of life that has been built across millennia of evolutionary processes and relations (Despret 2017). Further, as Dominique Lestel observes, when a species goes extinct, not only is the natural world impoverished but so is human life. He suggests that the ongoing loss of animal biodiversity represents a "psychological, intellectual, moral, and spiritual" (Lestel 2014: 320) impoverishment of humanity. When a species goes extinct, certain possibilities for co-creating new meanings, new relations, and new ways of being in the world go with it.

In view of the current crisis, animal studies theorists are developing new ways of understanding what is at stake with extinction as well as new strategies for trying to slow and even reverse present trends. In terms of learning to think more carefully about extinction, scholars like Thom van Dooren (2014) and Zoe Todd (2018) have urged us to think about extinction from what we might call the "underside" of our present moment, that is from the perspective of those animal beings and groups who are facing annihilation. For van Dooren, this entails familiarizing ourselves with particular species currently undergoing extinction and with unique individual members of those species. To this end, van Dooren looks specifically at albatrosses, vultures, whooping cranes, penguins, and crows and uses a detailed narrative and ethological approach in order to introduce readers into bird worlds and ways of life (which he calls "flight ways"). Immersed in these bird worlds, readers gain a richer sense for how birds live and die, and of what goes missing when their worlds and ways of life disappear. Writing from a decolonial perspective (see the entry on **colonialism** for more on this approach), Todd proposes considering species extinction in view of longstanding human–animal relations between Indigenous peoples and the animals with which they have lived and co-evolved for millennia. For Todd, human–fish relations in Indigenous lands in Canada provide an important site from which to rethink the stakes of extinction and for considering what survival might look like going forward. In the Indigenous communities Todd discusses, fish are seen as kin from whom human beings have much to learn about survival in difficult times, especially given the long evolutionary history of fish. Further, she suggests that Indigenous communities offer a number of compelling strategies for repairing and restoring relations with animals in an age when more and more of them are on the brink of disappearing.

In addition to such gestures of restoration and reparation, animal studies theorists and activists are considering other means of dealing with the ongoing extinction crisis. Options ranging from relocating species, to lifestyle changes that reduce ecological footprints, to more ambitious strategies such as **rewilding** have been proposed. More recently, technological interventions such as cloning, de-extinction, and genetic engineering are becoming live options, although many animal studies scholars are generally critical of these kinds of technological interventions (Oksanen and Siipi 2014).

Further reading: Chrulew and De Vos 2018; Heise 2016; Kolbert 2014; Parreñas 2018; Rose, van Dooren, and Chrulew 2017

FEMINISM

The term feminism refers to a wide variety of discourses, practices, and political struggles aimed at combatting sexism and achieving justice for women. This very general definition fails to do justice, however, to the rich history of feminist politics as well as to the several theoretical and activist "waves," or fundamental transformations, that feminism has undergone since its inception. This entry does not cover every aspect of that history or the entire range of feminist theories and practices that exist today. Rather, the focus here is on the most influential ways in which animal studies has been brought into dialogue with feminist concerns. In particular, emphasis is placed on work that has emerged over the past three decades, when interest in the connections between the status of animals and women markedly increased.

Pro-animal feminist theorists and activists have posited a number of connections among animals and women. While some theorists have noted the important role women have played in the genesis and continued growth of the **animal welfare and animal rights** movements (Gaarder 2011), the bulk of the material on the relations between women and animals has focused on common and overlapping forms of oppression and domination (Gruen 1993). At the core of these joint modes of oppression is a process that is sometimes referred to as "othering" (Gruen and Weil 2010). In societies that are speciesist and sexist, both women and animals are seen as "others," which is to say, they are viewed as belonging to the class of beings who are seen as other than, or not fully and properly, human. Such distinctions are made on the basis of the alleged absence or lesser degree of agency, rationality, and intelligence among animals and women. In de-humanizing women and sub-humanizing animals, the process of othering moves both groups into the class of entities that can be used as instruments and tools, thereby denying women and animals their individual uniqueness and agency. This "logic" of othering paves the way for numerous kinds of economic, political, and linguistic violence; and while women and animals do not necessarily suffer from precisely the same forms of exploitation, pro-animal feminists argue that the devaluation and objectification of both groups intersect in important ways (see **intersectionality**).

While the joint oppression of women and animals is widely acknowledged in animal studies today, there are several ways in which one might frame and challenge such oppression. A few of the more influential responses within animal studies include: (1) the care ethics approach, (2) the poststructuralist approach, and (3) the new feminist materialist approach. The *care ethics* approach to animal issues is grounded in research pioneered by feminist psychologist Carol Gilligan (1982), in which she contrasts an ethics of individual rights (typically associated with boys and men) with an ethics of caring relations (typically found among girls and women). The latter ethic is grounded in a sense of concern and responsibility for particular others with whom one is in relation, whereas the former ethics stresses the relative autonomy of individuals and aims to minimize mutual interference. Feminist animal care theorists extend Gilligan's care ethic framework to include animals within the scope of care and responsibility (Adams and Donovan 1995). In trying to restore more caring relations with animals, these pro-animal theorists suggest we should understand ourselves to be in dialogue with animals, and that discerning the best ethical course of action to take with a particular animal requires attention to the details of the relational contexts in which one finds oneself. Feminist animal care theorists generally believe that loving, emotional relationships with animals provide the best foundation from which to relate to animals and that cultivating such relationships is essential to overcoming the speciesist prejudices that characterize the dominant culture. This kind of emotion- and care-based ethical approach contrasts strongly with animal rights and **animal liberation** frameworks, both of which derive from highly rationalistic frameworks that emphasize justice and non-interference. Feminist animal care theorists argue that animal rights and animal liberation approaches tend to reproduce a bias in favor of rationality and autonomy (at the expense of emotion and relation) typical of sexist cultures. By placing care at the center of ethics, these theorists maintain that both feminism and animal studies will be positioned to move beyond the lingering speciesist and sexist dogmas that plague both traditions.

While the care ethic approach is the most fully developed form of feminism in animal studies, recent years have seen other feminist frameworks brought into dialogue with animal issues. One such approach, the *poststructuralist* approach (a name that derives from a trend in theory and philosophy that developed in view of and in response to structuralism), has used the work of feminist philosopher Judith Butler and related figures to think about how animals might come to matter in an ethical sense. Although Butler herself has not written extensively on animals, she has developed ways of thinking about how certain lives and deaths do or do not count in a given society—an issue that is of great interest both to feminists and to animal studies theorists. Butler uses the term "grievable" to refer to those persons whose lives and deaths matter (for example, quintessential human subjects) over against those who do not (for example, persons who have been dehumanized, including women, the poor, and people of color). The latter group find themselves in a position of heightened vulnerability and precariousness, and Butler's work aims at uncovering the linguistic, social, and institutional dynamics through which certain human beings end up in such conditions (Butler 2004; Butler 2009). As animal studies theorists have argued, however, animals are particularly illustrative instances of beings who occupy vulnerable and precarious positions in the social order and who have been deemed to fall outside the scope of grievability. As such, it is necessary to consider how processes of human marginalization also affect and involve the sub-humanization and de-valuation of animal life (Iveson 2012; Stanescu 2012); conversely, there is also an urgent need to consider the kinds of political interventions and affirmative practices that might allow animal lives to come to matter and animal deaths to become grievable (Jenkins 2012; Donaldson 2015).

New materialist feminists have also shown an interest in the connections between women and animals, although their approach differs considerably from both the care ethics and poststructuralist approaches. New materialist feminists are most concerned to overcome versions of feminism that have downplayed the importance of the physiological and biological materiality of women's bodies and, in a related manner, the materiality of animals and the natural world. While earlier versions of feminism (understandably) sought to demonstrate that women were, like men, fully rational beings with agency and autonomy, this view of liberal equality was often purchased at the expense of stripping animals, nature, and all aspects of human existence associated with these realms of value and spontaneity. New materialist feminists argue that it is high time for feminists to reconsider their reductive conception of nature, animals, and embodiment and to recognize that the material world is not fixed or static, passively awaiting human social constructions. Rather, animals, nature, and matter more generally should be conceived of as having agency in their own right, as being spontaneous, resistant, and vibrant (Barad 2007; Bennett 2010). In view of animals in particular, this kind of materialist approach goes beyond merely critiquing the common oppression of women and animals and moves toward the affirmative project of developing new kinds of relations with animals and new practices that affirm human and **more-than-human** materiality. Whether these new relations go by the name of **posthumanism** (Braidotti 2013) or **companion species** (Haraway 2003, 2008), materialist feminists aim to provide us with a different view of what human beings and animals might become, together and in relation with one another.

See also: agency; absent referent; carnophallogocentrism; dehumanization; empathy; ethics

Further reading: Adams and Donovan 1995; Alaimo and Hekman 2008; Birke and Holmber 2018; Donaldson 2015; Donovan and Adams 1996; Gruen and Weil 2010; Rossini 2006

HUMAN/ANIMAL DISTINCTION

The human/animal distinction refers to the numerous ways in which human beings have been differentiated from animals. For animal studies theorists, this distinction is an issue of central importance, as accounts of the differences between human beings and animals have a substantial impact on the status of animals within society. Throughout history and across cultures, there have been various approaches to thinking about differences and relations between human beings and animals (Descola 2013); but most work on the topic in animal studies takes as its critical point of departure the *dominant* approach to the human/animal distinction (Derrida 2008: 40). The dominant view has a long heritage behind it, constructed from philosophical, intellectual, and socio-economic trends that are global in nature but that originate largely in Western culture (see Preece 2011 for a contrasting view of Western attitudes toward animals). Although the dominant approach acknowledges that human beings and animals share certain traits and behaviors, it portrays human beings as having certain unique traits or markers that all other animal species lack. To identify these distinguishing markers of humanity is to pinpoint what is sometimes referred to as the "anthropological difference" (Glock 2012). There are a number of candidates for this anthropological difference, including: the ability to use **language**; awareness of one's own **death**; freedom and **agency**; the capacity to be motivated by ethical concerns; the capability to think in creative and rational ways, and so on. What makes this dominant approach to thinking about the human/animal distinction potentially questionable is not simply its tendency to draw sharp distinctions between human beings and animals (a theme that will be discussed more fully below), but its reliance upon a normative framework that suggests having such purportedly unique capacities should grant human beings higher ethical, legal, and political standing. Much of the work done in animal studies contests this normative premise and also challenges the dominant approach to distinguishing human beings from animals.

The first such challenge issues primarily from philosophers, legal theorists, and activists who try to establish a certain *identity* among human beings and animals. This approach maintains that, despite certain dissimilarities, there exist fundamental and evident similarities at ethical and legal levels between human beings and animals. Thus, even if we could discern a noticeable difference between human beings and animals in terms of, say, the capacity for symbolic language, such differences do not eliminate the fact that both human beings and animals are similar in a whole host of other ways. For instance, human beings and many (if not all) animals are: **sentient**, that is, they have the capacity to feel pain and pleasure (Singer 2001a);

conscious subjects who have a stake in how they are treated (Regan 1983); agents who have goals and desires they wish to fulfill (Cavalieri 2001); and persons who have intentionality and a sense of self (Wise 2000). Although it is true that not every being designated as an "animal" has all of these capacities, many do; and these shared capacities should, in principle, have thoroughgoing implications for how those animals are treated. A standard principle of legal and ethical reasoning is that relevantly similar beings should receive similar consideration. It would seem, then, that animals who demonstrate they have ethically and legally relevant abilities and capacities should be granted **equal consideration** in terms of their ethical and legal interests. Such consideration does not, of course, mean that animals will receive identical treatment to human beings under every circumstance (for instance, many animals would presumably not be interested in equal educational opportunities); but where animals do have ethically relevant interests (say, in not being harmed or not being held captive), it would seem arbitrary to ignore them.

Another influential approach to thinking about the human/animal distinction derives from thinkers and activists who seek to refine and complicate what *difference* means. The dominant approach, as noted, tends to think of human beings and animals as sharply differentiated and opposed along the lines of a single trait or capacity. From the perspective of the difference approach, this simple, oppositional way of drawing lines between human beings and animals reduces the rich plurality of both groups. Sharp oppositions and distinctions of this sort tend to make us think that both sets—human beings on the one hand, animals on the other—are homogeneous in nature. Thus, rather than seeing animals in terms of their overabundant differences and singularities at both the level of individuals and species, the dominant schema would have us encapsulate all animals within a single category. Likewise, if we approach human beings using this kind of oppositional and reductive thinking, we will fail to notice all kinds of relevant and important distinctions that might be made among human beings. The difference-based approach, then, suggests that the traditional human/animal distinction is too flat and one-dimensional, and that we should attend to the complexity of differences between, within, and among human beings and animals. Further, for most thinkers and activists who adopt this approach, such differences are meant to be affirmed and even celebrated, in contrast to the dominant approach which tends to devalue any being or group that differs from paradigm-case human beings. In this refined account of what constitutes human and animal differences, there is no compelling reason to grant higher ethical rank to one group of beings over another. This approach has been enormously influential within animal studies, and variants of it can be found especially among those figures influenced by the work of Jacques Derrida as well as Carol Adams, Josephine Donovan, and other feminist theorists.

The two approaches mentioned so far have been crucial for calling attention to the limits of traditional ways of thinking about human beings and animals, but they strike some animal studies theorists as inadequate in fundamental ways. These limits have led to a third approach that focuses on the manner in which human beings and animals are *indistinct* or *indiscernible* (terms associated, respectively, with Agamben 2004 and Deleuze and Guattari 1987). This alternative approach argues that it is not simply the case that animals more or less resemble paradigm-case human beings (as the identity approach suggests); rather, even paradigm-case human beings participate in surprising and profound ways in shared aspects of animal life (for example, in being embodied, in dying, in affectivity, and so on). And while it is true that there are innumerable differences between, among, and within human and animal existence (as the difference approach suggests), this third approach suggests that genuinely attending to those differences tends to undermine altogether the project of determining anthropological differences. In other words, as we learn more about the diversity and richness of animal life, it no longer seems necessary or plausible to continue the search to determine human uniqueness over and against animals as a whole. If that project of establishing human uniqueness is set aside, then other possibilities for thinking, living, and relating open up that have traditionally been overlooked. Such alternative ways of being in the world might even serve to make us rethink altogether what it means to be human and animal, which is a core impetus behind such concepts as **becoming-animal** (associated with Deleuze and Guattari) and **companion species** (associated with Donna Haraway). This third approach resonates in many ways with key trends in **posthumanism**, which also proceeds under the assumption that traditional beliefs about the uniqueness of human beings are no longer tenable.

See also: animot; ethics; feminism

Further reading: Calarco 2015; Ohrem and Bartosch 2017; Steiner 2005

HUNTING

The term hunting comprises a number of ways in which animals are tracked and killed. Hunters kill animals for sport, for subsistence, as trophies, to manage the ecology, to eliminate pests, for profit, and even for entertainment. Alongside living with **pets** and watching **wildlife**, hunting is one of the more common ways in which human beings come to know and interact with animals. In recent years, the number of people hunting has declined, but there are still considerable numbers engaged in the activity. The United States Fish and Wildlife Service reports that, in 2016, 11.5 million people undertook some form of hunting activity (excluding fishing) (U.S. Department of the Interior et al. 2016). One prominent defender of hunting estimates that more than 200 million animals are hunted per year in the United States (Swan 1995), but this figure does not include the many more millions of hunted animals that go unreported. (There are no reliable estimates for the number of animals hunted globally on an annual basis.) Among the most commonly hunted animals are big game animals such as deer and bighorn sheep, small game such as squirrels and rabbits, and birds such as doves.

Of the various kinds of hunting, *sport* or *recreational* hunting in particular has been the most hotly debated in the academic literature and in popular culture. Defined as hunting animals for recreational purposes, sport hunting is typically defended on the grounds that it is ecologically beneficial. Without sport hunters culling and controlling certain populations of animals, it is argued, certain species that lack sufficient predators will overpopulate a given bioregion. Furthermore, sport hunters argue that they are the primary supporters of wilderness conservation, given that wilderness provides the resources for their preferred mode of recreation. Some philosophers and ecologists, such as José Ortega y Gasset and Aldo Leopold, also defend recreational hunting by appealing to its spiritual properties. From this perspective, sport hunting is seen as a means of reconstituting a primal human connection with wilderness and providing a means of engaging with animals outside of cultural settings (see Causey 1989 and Kowalsky 2010 for fuller discussions and defenses of sport hunting of this sort).

Critics of sport hunting remain unpersuaded by these points. They argue that a balanced and healthy ecology can be achieved through means other than hunting; that wildlife and wilderness conservation can be decoupled from hunting and still be sustained; and that the spiritual benefits of hunting do not outweigh the pain and suffering that hunting causes animals (King 1991). Some animal ethicists argue, in addition, that hunting reinforces troubling patriarchal attitudes (Kheel 1996) and is logically inconsistent with its own ethical codes (Luke 1997). With regard to the latter point, if sport hunting proceeds from the premise that animals have intrinsic value and that they should be killed respectfully (as most sport hunters would insist), then it seems that such ethical recognition could best be achieved by avoiding hunting and killing animals altogether—especially in those instances where such hunting is unnecessary.

Of course, there are instances where the case could be made that hunting is necessary, namely, in regard to *subsistence* hunting. If animals are hunted by human beings and their flesh is eaten in order to meet vital nutritional needs, some theorists argue that such a practice is far less ethically questionable (Wenzel 1991; see Donner 1997 for an opposing view). Further questions about subsistence hunting arise when it is carried out by Indigenous peoples who hunt both for subsistence and as part of continuing or reconstituting traditional ways of life that have been interrupted by **colonialism** (Reo and Whyte 2012). (Readers interested in exploring these issues further might consult the work of ecofeminist Greta Gaard [2001], who focuses on the complex ethical and political questions surrounding hunting and colonialism in the context of the recent Makah whale hunt; see also the film by Arnaquq-Baril 2016).

The bulk of the debates about hunting to date have been focused on ethical concerns about the suffering and death of individual animals who are hunted and the weight such concerns should carry against the interests of hunters and hunting-based communities. In the coming decades, as problems stemming from **climate change**, ecological degradation, and biodiversity loss become increasingly prominent, the debates over hunting will undoubtedly be altered to take this changing ecological context into account.

INTERSECTIONALITY

The concept of intersectionality refers to the multi-layered vectors of power that shape an individual's subjectivity (that is, one's sense of self and identity). Although the term is now used across a wide variety of disciplinary and activist contexts, it has its historical and intellectual origins in debates within feminist theory and politics in which Black and other women of color feminists (sometimes referred to as "womanists") argued that, despite the importance and power of traditional feminist critiques of sexism, traditional **feminism** needed to be supplemented by additional perspectives and critical standpoints such as **race** and class. Women of color feminists suggested that in dealing with political issues, they were often unable to choose gender *or* race *or* class alone as the salient vector of analysis insofar as all of these axes of power intersected in their experiences of oppression.

As these debates developed into the late 1980s, legal theorist Kimberlé Crenshaw (1989) used the term "intersectional" to capture the underlying logic of this womanist critique of classical feminism. Crenshaw pushed back against what she called "single-axis" frameworks that place gender, race, or class alone at the center of analysis. In some cases, especially in the case of Black women, all of these axes are often relevant. Consequently, Crenshaw argued that critical theories should, in principle, be intersectional and attend to the ways in which power and oppression tend to interlock and intersect across multiple lines. As the concept of intersectionality was expanded and put to work in these additional domains, the various axes of power that have come under critical scrutiny have similarly multiplied. Thus, even though early writings on intersectionality tended to focus on the central sociological categories of analysis (namely, race, class, and gender), intersectional analysis today has been expanded to include such categories as sexual identity, disability, coloniality and a whole host of additional forms of power and domination.

For many animal studies theorists, intersectionality is simultaneously a promising and problematic concept when applied to animals. In order to understand intersectionality in a more capacious and less human-focused manner, some theorists have suggested that **speciesism** should belong to the list of intersectional axes. In this vein, Andrzejewski, Pedersen, and Wicklund (2009) provide a helpful way to think about the links between the mistreatment of animals and intersecting forms of social injustice. They argue that speciesism has deep historical and structural connections with other forms of social injustice, and that exploring these connections can (1) help clarify the way such injustices mutually reinforce one another, and also (2) build connections among otherwise disparate and seemingly incommensurable justice struggles. Following their lead, we might link speciesism with sexism, racism, classism, and heteronormativity in some of the following ways:

- With regard to speciesism and *sexism*, literature from feminist and ecofeminist theorists highlights the overlapping modes of oppression experienced by both women and animals. In sexist, meat-eating societies, women and animals are often referred to in terms that de-individualize and fragment their uniqueness; and exploited animals and women are often sexualized and characterized as passive beings who welcome their exploitation.

- In view of speciesism and *racism*, many of the arguments used in defense of mistreating animals have been used to justify slavery and other forms of racial injustice. Similarly, the technologies and methods developed for the mass killing of animals have been deployed in racist mass killing of human beings (for example, in the Jewish Holocaust). Important parallels between racism and speciesism can also be found in institutions such as prisons and **zoos** (where both marginalized human beings and animals have been captured and confined against their will) and practices such as medical **experimentation** (in which both animals and marginalized human beings have undergone involuntary and painful experiments to advance scientific knowledge).

- In regard to speciesism and *classism*, the exploitation of animals has been central to the development of privileged economic classes. In addition, animals and economically dispossessed humans have both been used as mere means to the end of profit maximization. In important ways, both human beings and animals belong to the working class, and the liberation of workers can be seen as tied up in important ways with the liberation of animals.

- In terms of the relations between speciesism and *heteronormativity*, there are important connections between the normalization of heterosexuality and the normalization of meat eating. Both practices, while contingent and far from being universally practiced, masquerade as universal and timeless and thereby make the current social order appear fixed and unchangeable. Further, the bullying and violence that is often visited on those who assume alternative sexual and gender identities can be seen to run in parallel with the abuse of vulnerable animals.

This is but a brief list of the ways in which traditional, intra-human, intersectional axes of analysis might be seen to intersect and interlock with power structures that affect the lives of animals. As promising and insightful as such intersectional work might be, from the perspective of animal studies this approach also has certain limits in terms of its explanatory reach. It is commonly agreed that intersectional analyses allow for the autonomy of each region of analysis and aim to avoid the kind of reductionism in which one form of power might be explained away by another (for instance, gender discrimination cannot be reduced to class-based economic exploitation, even though there are often deep connections between the two). If this kind of autonomy is borne in mind, then it would be equally important to note that the exploitation of animals must be understood as having its own specificity and importance, independent of any contingent linkages with human injustice. Thus, the exploitation of animals should be seen as an important issue in its own right and should not become a point of concern *only* insofar as it happens to be linked to one or another form of human social injustice (Clark 2012). Given that animal issues are often not taken to be of equal importance in human social justice contexts, this point will no doubt be a source of contention as intersectional analyses that emphasize speciesism develop in the coming years.

Similarly, there are limitations with an intersectional approach in regard to animals to the extent that intersectional thinking was originally developed to understand the formation of marginalized human subjectivity in the context of human social institutions. While one of the key lessons of animal intersectionality theorists has been that such vectors of intra-human power relations bleed into the animal world, it is not altogether clear that the **captivity** and confinement of animals has been carried out primarily with the aim of forming animals (or other nonhuman beings) into "subjects" with specific "identities." Power works in various ways and toward various ends; and to understand how power circulates through and among animals in particular might require different analytical approaches that focus specifically on animals and that attend to alternative kinds of social formations (Puar 2011; J. K. Stanescu 2013).

See also: absent referent; capitalism; captivity; feminism; queer; race; work

Further reading: Adams and Gruen 2014; Cudworth 2014; Dinker and Pedersen 2016; Glasser 2015; Hancock 2016; Harper 2010; Lykke 2009; N. Taylor 2013; Twine 2010

INVASIVE SPECIES

Invasive species are species that have been introduced by human beings into regions beyond their typical geographic range and, once introduced, expand their numbers by displacing species native to that habitat. In recent years, studying the effects of invasive species and developing strategies for adapting to or entirely eradicating such species has spawned an entire sub-discipline of biology known as *invasion biology*. The chief concern in this field is that invasive species can initiate processes (such as altering nutrient cycles and degrading habitats) that lead to dramatic biodiversity loss and even species **extinction** under certain circumstances (Simberloff 2011). Such losses are usually characterized as being economic in nature (the perspective typically adopted by wildlife resource managers) or ethical and aesthetic (the perspective most common to environmentalists).

The science surrounding invasive species is not without its detractors. Some scientists suggest that, while novel species often do significantly alter existing ecosystems, there is little evidence supporting the premise that they pose "an apocalyptic threat to biodiversity" (Davis et al. 2011; see also Sagoff 2005). Hence, rather than an uncritical endorsement of invasive species removal or a wholesale rejection of all such projects, these critics argue that strategies concerning how to deal with novel species must be developed on a case-by-case basis and without the implicit assumption that native species are necessarily harmful to ecosystem health (Chew and Carroll 2011). In response, invasion biologists maintain that the problems posed by invasive species are, in fact, among the main conservation threats facing us today and that we should rapidly pursue a wide array of management strategies and eradication technologies (Simberloff 2010). Likewise, some animal studies theorists, such as Donna Haraway, endorse the idea that invasive species are a serious problem and that their non-native status offers persuasive reasons for their being killed and even eaten (Haraway 2008; see V. Stanescu 2016 for a fuller discussion).

On the whole, though, animal studies theorists have tended to be critical of the notion of invasive species and have raised a number of objections to the discourses and practices surrounding the concept. James Stanescu and Kevin Cummings (2016a) argue that invasion biology employs a "logic of extermination" that needs to be interrogated in view of its tendency to render invasive species killable with impunity. Further, they argue that, at a historical moment when important strides are just beginning to be made toward better ethical treatment of animals, it is important not to cede ground to policies that would treat animals with such striking disrespect. Other scholars like Maneesha Deckha and Erin Pritchard (2016) suggest the discourse on invasive species gains much of its economic and social support by piggybacking on xenophobic ideas about the respective value of human natives (portrayed as having higher value and essentially belonging to a given place) and human immigrants (portrayed as having lower value and being invaders of a given region). In addition, they argue that labelling a particular species "invasive" tends to reinforce **anthropocentrism** and pernicious ideas about animals being fundamentally lower in worth than (certain groups of) human beings. A more ethical approach to such species, they counsel, is to learn to co-exist with them and see them as integral members of human and natural communities.

Other scholars have suggested that in an era of rapid ecological change and shifting geographical ranges (an era scholars often refer to as the **Anthropocene**), increasing numbers of species will be dislocated from their traditional geographical ranges, to the point where trying to discern which species are native or invasive will make little sense (Hill and Hadly 2018). On this account, with the onset of the Anthropocene we have entered a "post-native" era in which the goals of traditional conservation biology will have to be rethought in fundamental ways. One might also note that, in such times of rapid **climate change** and geographical transformation, it is difficult to determine with any certainty which kinds of non-traditional species will actually harm the responsiveness, health, and plasticity of a given ecosystem or bioregion; in fact, some newly arrived species might actually *improve* the long-term health of a particular region or ecosystem (Lorimer 2015). As species of all sorts are becoming increasingly entangled on a warming and rapidly changing planet, difficult questions about how to deal with these unprecedented forms of bio-cultural contact will become a pressing issue both for specialists in animal studies and for the public as a whole.

Further reading: Coates 2007; Nagy and Johnson 2013; Simberloff 2013; Stanescu and Cummings 2016b.

LANGUAGE

Of the various markers used to distinguish human beings from animals, language is perhaps the most enduring and ubiquitous. In the Western intellectual tradition in particular, the capacity for language has consistently been understood to constitute the decisive "anthropological difference" (see the entry on **human/animal distinction**) from antiquity to the present. Whereas such capacities and characteristics as **mind**, understanding of **death**, tool use, and **agency** are now widely accepted as existing in at least rudimentary form among some animals, the ability to use language is still thought by many people (in both academic and non-academic settings) to belong exclusively to human beings. Recent findings in the field of animal language research are, however, complicating this traditional picture, if not entirely displacing it.

No one who studies animals in a sustained manner can fail to be impressed by the sheer complexity and variety of their modes of communication, which range from the remarkable waggle dance of honeybees to the beautiful songs of musician wrens. In order to navigate their environments, reproduce, find food, defend themselves against predators, establish kinship ties, **play**, and resolve conflicts, animals have developed a vast repertoire of communication techniques, including vocalizations, signaling behaviors, facial expressions, and bodily gestures (see Bradbury and Vehrencamp 2011 for a helpful overview of animal communication). The key question for researchers, though, is whether any of these communication systems rise to the level of the kind of language that we see in nearly all human societies. Language in this more refined sense goes beyond the fixed range of instinctual forms of communication that many animals use and involves the ability to employ language symbolically to refer to objects that are not present, to use and combine words in novel ways, to create new words for distinct objects and persons, and to employ complex grammar and syntax. Animal language skeptics argue that either some or all of these characteristics of human language are absent in the animal kingdom (Pinker 1994; Hauser, Chomsky, and Fitch 2002).

As a means of testing standard claims about the uniqueness of human language, scientists have devised several long-term experiments to see whether animals can, in fact, be taught human language. The two best known instances of such experiments are those involving an African grey parrot named Alex and a gorilla named Koko. In the case of Alex, who could mimic the sounds of human language, his trainer Irene Pepperberg claims that he learned in excess of 100 different words, could identify distinct objects according to several rubrics, and was able to communicate his inner states (Pepperberg 2009). Koko, who was generally unable to make human-like vocal utterances, was taught American Sign Language and was able to use some 1,000 signs and understand 2,000 spoken words (Gold and Watson

2018). Similar experiments have been done with other animal species, from dolphins to dogs, with varying levels of success (see Hillix and Rumbaugh 2004 for an overview of this research). Although some of these experiments demonstrate a striking ability among certain animals to learn a language that is not their own, even the staunchest defenders of animal language capacities do not believe the studies show animals reaching the level of language use characteristic of a competent adult human speaker.

Perhaps, though, this is the wrong approach to take when considering the relationship between animals and language. It might make better sense to learn the ways in which animals communicate among themselves in their own terms and see if those modes of communication rise to the level of language. This is what many ethologists have been doing for the past several decades with a number of different species, and the findings have been nothing less than astonishing. For example, it has been observed that vervet monkeys are able to make several distinct calls for different kinds of predators and that these calls generate distinct, correlated responses from their troop (if a call for an eagle is given, they look to the sky; if a call for a snake is given, they look down to the ground, and so on; see Seyfarth, Cheney, and Marler 1980 for the classic study on this topic). Extensive study of prairie dogs by biologist Con Slobodchikoff and his colleagues has shown similar capacities among this species for making distinct calls for different predators. Further, Slobodchikoff's research has found that prairie dogs are able to invent new words, have their own regional dialects, and are able to communicate such fine-grained details as the speed and size of approaching predators (Slobodchikoff, Perla, and Verdolin 2009). In another remarkable study, Stephanie King and colleagues demonstrated that bottlenose dolphins naturally develop distinct individual "signature whistles" that function much the same way human names do to distinguish one individual from another. This ability to name and differentiate individuals within a group is thought to be especially important for animals like dolphins who live in complex social environments (King et al. 2013).

The examples could be multiplied here, but the larger point is that this ongoing research suggests the gap between human and animal language is not as insuperable as it is often assumed to be. To be sure, no researcher is trying to make the case that human and animal languages are *identical* in every respect; rather, what emerging evidence implies is that the building blocks of human language are present in varying degrees across a wide variety of animal species. Even if human language is highly sophisticated and noticeably different from most animal languages, it appears that the differences here are (as in most other instances) a matter of degree rather than kind (Lents 2016: 273). Finally, it should be noted that this exercise in trying to determine whether animals can learn human language or whether animal communication rises to the level of human language tends to reinforce the very kind of **anthropocentrism** that most animal studies practitioners hope to undercut. To approach animals in view of determining whether they measure up to some predetermined human standard is to miss having a genuine encounter with other animals in all of their difference and richness. The question that most animal studies scholars would pose in this regard, then, is not whether animal communication is identical to human language but whether we might learn to see, think, and live differently by immersing ourselves in the complex and meaningful languages of animal others.

Further reading: Evans 2014; Pepperberg 1999; Radick 2007; Slobodchikoff 2012

LAW

Law is one of the central domains in which advocates seek protections for animals. There is an extensive and complex history of laws protecting animals, stretching back to antiquity and across several cultures. Legal advocacy for animals has in recent years become a viable and increasingly influential field of study and practice, with many leading law schools offering courses, certificates, and even subspecializations in animal law (for an overview of the rise of contemporary legal advocacy for animals, see Tischler 2008 and 2012).

Lawyers employ the legal system primarily to protect animals from cruel treatment. This work is carried out in three chief ways: (1) prosecuting animal abusers who violate existing animal welfare laws (for example, prosecuting pet owners who neglect their pets); (2) creating new laws to protect animals where protections are lacking (for example, the establishment of the Humane Slaughter Act, which seeks to make slaughtering processes as painless as possible); and (3) challenging existing laws that allow for animal harm and abuse (for example, banning private establishments from keeping animals in captivity for the purposes of **entertainment**). Among the more prominent groups who engage in this kind of legal advocacy is Animal Legal Defense Fund, which was founded in 1979 and has filed a number of prominent lawsuits in defense of and on behalf of animals.

One of the limitations of this way of using the law to advocate for animals is that it proceeds largely from within longstanding legal frameworks that treat animals as property rather than as persons. Thus, no matter how many animal welfare protections are established and violations are prosecuted, animals who are classified as property are ultimately viewed and treated by the legal system as things to be owned by human persons (Francione 1995). Some animal law experts argue that the only way to render law truly effective for protecting animals' interests is to fight for full legal personhood for animals. Advocates for animal personhood are well aware that achieving this goal is a long-term, uphill battle. They believe, however, that scientific evidence persuasively supports the conclusion that many animal species have the requisite capacities (such as autonomy, self-awareness, and **sentience**) for legal personhood (Wise 2005), and that consistent reasoning requires that we extend **equal consideration** to such species (Wise 2000). The implications of certain animals being granted legal personhood are far-reaching. If a given animal species comes to be seen as having full legal personhood, then members of those species could no longer be, for example, legally killed and eaten, held captive, forced to **work**, or used for entertainment against their will; moreover, their natural habitats would have to be protected, which would entail serious changes in dominant economic and agricultural policies. The most prominent organization dedicated to achieving the legal personhood of animals is the Nonhuman Rights Project, which was founded in 2007 by Steven Wise, an influential animal rights author and advocate.

For many animal studies theorists, the efforts by animal lawyers to extend legal protections and standing to animals are seen as a welcome and worthwhile endeavor. However, theorists such as Jacques Derrida (Derrida and Roudinesco 2004) and Cary Wolfe (2013) warn that over-reliance on the legal system for animal advocacy has potentially pernicious consequences. Given that animals have long been excluded from legal standing because they were not seen as full persons, the mere demonstration that *some* animals technically do qualify as full persons does little to challenge the exclusionary nature of person-based approaches to law. There will always be animals (and other human and nonhuman beings) who fail to meet the standards of full personhood and who will thereby be situated in a precarious position in relation to the dominant normative and legal orders. This critical perspective raises the question of whether the legal system can be adequately reformed to do justice to the lives of the majority of animals and other **more-than-human** beings. (For more on the limitations of dominant approaches to animal law from an animal studies perspective, see Otomo and Mussawir 2013.)

Further reading: Favre 2011; Schaffner 2011; Waisman, Frasch, and Wagman 2014

MEAT

In the context of animal studies, meat typically refers to animal flesh intended for human consumption. Meat is a controversial term, though, because as an abstract and general concept it glosses over the individual animal lives and complex historical, social, and ecological factors that make commodified meat possible (Fiddes 1991). Before exploring these obscured realities, it is important to survey some of the basic facts about the consumption of animal flesh in contemporary societies. Global consumption of meat has markedly increased over the past five decades, from roughly 70 metric tons per year in 1961 to 330 metric tons in 2017. What is most remarkable about this upsurge is not simply the increase in the sheer amount of meat consumed, but the acceleration of the rate of *per capita* meat consumption. The average amount of meat consumed per individual per year has more than doubled during this time, increasing from some 20 kilograms of meat per year in 1961 to nearly 45 kilograms per year today (OECD/FAO 2018; see also Ritchie and Roser 2019).

This upward trend is striking, and it points toward a fundamental transition in the content of human diets, especially among wealthier inhabitants of advanced industrialized nations such as the United States and Australia and in rapidly developing countries such as Brazil and China. Political ecologist Tony Weis (2013) uses the term *meatification* to describe this change in global food consumption, in which meat has moved from the periphery of human diets to their center. Weis notes that by focusing specifically on the increase in meat consumption (rather than, say, increased "protein" consumption), we can begin to remove the ideological veneer that covers our eating practices and take up the task of thinking more carefully about what meat is and the complex socio-economic systems that enable its mass production. In line with this aim, a number of scholars have proposed concepts that highlight the overwhelming presence of animal flesh in global consumption patterns, including such terms as **animal–industrial complex**, *meat culture* (Potts 2016), and *carnism* (Joy 2011).

Whatever term is used, critics of this system raise a number of powerful objections against contemporary methods of meat production. They argue that high levels of meat consumption are: a leading cause of a number of human health concerns; a major contributor to a range of ecological issues, from water scarcity to pollution to **climate change**; a source of economic inequality in the poorer regions of the world where animals are raised; and a contributing factor to the food systems that mistreat those who work with meat animals, especially slaughterhouse workers who tend to be poor and live in economically and socially precarious conditions. Finally, critics argue that current systems of meat production (such as are found in **CAFO**s) treat animals themselves in manifestly unethical ways. The current debate over eating meat, then, cannot simply be reduced to competing positions over taste preferences or personal dietary choices. Critics of meat consumption are making the case that the economic and agricultural systems involved in the mass production of animal flesh for human consumption arise not just from consumer preferences but from a complex set of social ideologies and economic institutions that have myriad negative effects and that routinely strip animals of their dignity by reducing them to edible beings.

Underlying this critique of the dominant meat culture is the ethical conviction that animals ought to be treated more like human beings (or, at least more like human beings who have full ethical and legal standing). By raising animals to the level of human beings and re-moving them from the sphere of edibility, it is assumed that the injustices associated with meat culture will be drastically reduced and the door will be opened to developing healthier relationships with animals. But critics such as ecofeminist philosopher Val Plumwood (2002) and animal philosopher Dominique Lestel (2016) argue that efforts to remove animals entirely from the realm of social and **ontological** edibility stem from a myopic and fundamentally non-ecological view of animal life. From their perspective, an animal's being eaten is, ecologically speaking, an important part of the cycling of energy and nutrients through ecosystems—and to think that animals can somehow be saved from this fate is a dangerous illusion.

Does it follow from this way of thinking, then, that human beings should continue to eat meat unaffected by the profoundly unethical circumstances under which most meat is produced today? For Plumwood and Lestel, such a conclusion would miss the point. At issue is not defending contemporary meat culture; rather, what is at stake is recognizing that human–animal relationships will not change for the better until (1) human beings understand themselves to be fully animal, and (2) human beings see both themselves and animals as fated to being meat for other beings. The ethical aim from this perspective is thus not to raise animals out of the sphere of edibility and into the protected sphere of non-edible humans, but for human beings to recognize that they, too, are like animals insofar as they are also edible and belong to ecological cycles and trophic chains. Once human beings fully acknowledge that they are like animals and are always potentially reducible to being meat for another animal, then a deeper conversation about what it means to eat ethically and respectfully can be initiated. This approach to thinking about the joint edibility of human beings and animals could conceivably issue in either vegetarianism (Plumwood) or ethical carnivorism (Lestel).

Debates regarding the ethics, politics, and even ontology of meat are currently changing shape in dramatic ways, due in large part to technological developments in food production. Driven primarily by concerns with the health, ecological, and animal welfare issues related to current modes of meat production, a number of commercial meat substitutes and alternatives are being developed. The two main meat alternatives are *plant-based meat substitutes* and *in vitro meat* (also called "cultured," "lab-grown," or "clean" meat). Plant-based substitutes have been available for several decades, but in the past few years, substitutes that are nearly indistinguishable in taste from animal flesh have become widely available in grocery stores and fast-food restaurants. These alternatives generally carry a much lower carbon footprint than factory-farmed meat and eliminate many of the animal welfare concerns of commercial animal agriculture. However, given that these products are produced on a mass scale and circulate through global market supply chains, they still confront challenges with being humane and sustainable in view of their shipping and transportation and in terms of producing the plant materials needed to scale these products for national and global markets (Goldstein et al. 2017).

In vitro meat products face a number of regulatory and developmental hurdles and are not yet widely available, but they have garnered intense interest and robust financial support from venture capitalists and some animal welfare advocates who believe that these alternatives will eventually meet global meat demand with far fewer of the negative side-effects of conventional production methods (Shapiro 2018). What makes in vitro meat controversial among some animal ethicists and animal studies scholars is that, despite drastically reducing the direct harms to animals, it is ultimately an animal product (in vitro meat is derived from animal cells and is currently grown in most cases using fetal bovine serum). In contrast to plant-based alternatives, in vitro meat thus does not ask consumers to rethink or modify their commodified relationships with animals; instead, it simply changes the methods in which meat is grown. Thus, even if in vitro meat can eventually be produced without causing animals direct harm in the process, it still maintains a largely anthropocentric attitude toward animals and the more-than-human world as a whole, an attitude that will most certainly persist in a host of other human–animal relationships (see Alvaro 2019 and Miller 2012 for further reflections on this point).

See also: absent referent; Anthropocene; anthropocentrism; carnophallogocentrism; ethics; vegetarianism and veganism

Further reading: Donaldson and Carter 2016; Emel and Neo 2015; Weis 2017; Winders and Ransom 2019; Wurgaft 2019

MIND

The question of whether and to what extent animals have minds comprises several complex empirical and philosophical issues. Are animals aware of themselves? Do they have their own perspectives? Are animals aware of others *as* others? Can they take another's perspective? Are animals able to respond in novel and creative ways to challenges encountered in their environments, or are they limited to pre-determined instinctual reactions? These are just a handful of the questions that occupy contemporary researchers who focus on animal minds and they form the main focus of this entry.

The notion that animals might have minds was not seriously entertained by most scientists until just a few decades ago. Prior to this time, "Morgan's canon" (from psychologist C. Lloyd Morgan), which stipulates that a behavior should never be explained using a higher psychological faculty if it can be attributed to one that is more basic, ruled the fields of psychology and zoology. In practice, this led to most scientists arguing that animal behavior could be adequately explained by non-conscious instincts and without recourse to more sophisticated faculties such as mind, consciousness, or awareness. With the rise of behaviorism, this general denial of animal minds became the dominant paradigm, and researchers who dared to suggest animals had minds risked being dismissed for committing the "sin" of **anthropomorphism**.

In the 1970s and 1980s, however, highly regarded animal researcher Donald R. Griffin began making the case that the dismissal of animal minds was premature and that there were good evolutionary reasons for believing that the mental and emotional lives of human beings and animals are continuous. Rather than maintaining that any one action or behavior proved that animals had minds, Griffin's work suggested that there were several behaviors and capacities (such as coordinated hunting and complex forms of sociality) characteristic of some animal species that implied varying levels of conscious and responsive awareness. Griffin's writings helped birth a research paradigm called *cognitive ethology*, which builds on classical **ethology** to include a careful examination of animal minds as well as the inner, subjective mental states of animals (see Griffin 1976, 1984, and 1992).

When exploring the issue of animal minds, one of the first things to consider is whether animals are aware of themselves as individuals and can differentiate themselves from other individuals. In human psychology, the standard test for determining such matters is the *Mirror Self-Recognition* test. In this test, a mark is placed on the body of the subject in a place that cannot be seen except in a mirror. To pass the test, the subject must notice, explore, and try to remove the mark. Human children usually pass this test by two years of age; and it was long thought that animals were generally unable to pass it at all. However, in recent years, variants of this test have been given to a number of animals, and several have successfully passed it, including dolphins, elephants, great apes, and magpies (Gregg 2013). Although passing this test generally indicates a fairly high level of self-awareness, critics argue that with animals it might only be an indicator of bodily awareness and not of any deep, reflective understanding of oneself or others (for a helpful collection of essays on this contentious topic, see Parker, Mitchell, and Boccia 1994).

In order to determine whether animals do, in fact, have this more robust sense of self, it would be useful to know whether they can distinguish between their own state of mind and that of others. In other words, we might be able to get a better sense of whether animals have their own perspective by seeing if they are able to infer that others have their own perspectives. This ability to engage in varied self-and-other perspective taking is referred to by scientists as having a *theory of mind*. The earliest experiments to test whether animals have a theory of mind were performed with chimpanzees (Premack and Woodruff 1978), and the results were impressive but did not entirely convince skeptics. Since then, scientists have gathered evidence from a variety of experiments with wild and laboratory animals that seem to have determined at least some animals have a theory of mind (see Balcombe 2010 for an overview of this research). Perhaps the most persuasive experiments on this question are those that employ anticipatory looking tests to track the gaze of animals as they try to infer where others will look for objects; in some cases, animals are not only able to anticipate another individual's actions but can even anticipate another individual's *false* beliefs (see Krupenye et al. 2016 for an experiment of this sort on three different species of great apes; see also *Science Magazine* 2016 for video of the experiment).

Another indicator of having a mind is the ability to respond to one's environment in novel ways that go beyond hard-wired, instinctual reactions or conditioned behaviors. This ability, which is most prominently on display in acts of problem-solving, is often considered to be exclusively human; but here too the evidence suggests that animals have minds capable of interacting with and modifying their environment in novel ways. If we look at *tool use*, for example, we find an incredible variety of species, from mammals and birds to cetaceans and invertebrates, who use tools in creative and ingenious ways (see the essays in Sanz, Call, and Boesch 2013 for a helpful overview). What is more, in a handful of species, novel use of tools is passed along from one generation to the next, implying a kind of cultural learning among some animals (Whiten and Mesoudi 2008). These kinds of creative and inventive activities imply that mind is present insofar as they show a given animal has some novel end or purpose toward which it is striving and is capable of creating and executing a plan over time to meet that end.

As mentioned above, these topics represent only a few of the issues that belong to the field of animal mind research (although they are, arguably, the core research questions). Other important questions concern whether animal communication rises to the level of **language**, whether animals have moral **agency**, and what sorts of memory animals might have. In general, the evidence unearthed by the recent scientific work on animal minds has led many scientists and the public at large to adopt a more generous and charitable view of the intelligence of animals. Thus, even though human beings display particularly impressive levels of cognition, consciousness, and self-and-other awareness, it has become increasingly untenable to maintain that no other animals possess these capacities and abilities in at least some degree.

We should note in closing one of the paradoxical implications of this research paradigm. Insofar as experiments on animal minds have tended to demonstrate that many animals are quite sophisticated, sensitive, and responsive creatures, these results call into question the legitimacy of the continued development of such experiments—especially when they are carried out on animals who are held in **captivity**. Many of the most persuasive and well-designed laboratory experiments on animal cognition place subject animals in positions of deep vulnerability, cause them evident frustration and irritation, and further cement the notion that animals are objects of research rather than subjects of a life. For these reasons, the scientific study of animal minds stands at a crossroads. Will the field continue to subject animals to captivity and forced **experimentation** in order to obtain verifiable and repeatable scientific evidence? Or will it turn in a more ethological direction and study free animals in their native environments, with all of the unpredictability and ambiguity intrinsic to such research?

MORE-THAN-HUMAN

The adjectival phrase "more-than-human" derives from the work of David Abram and appears in the subtitle of his influential book, *The Spell of the Sensuous: Perception and Language in a More-than-Human World* (1996). The term is used by Abram to describe animals and nature in a way that avoids the denigrating language of "nonhuman" or "subhuman," terms that suggest animals and nature are less than, or inferior to, human beings in some fundamental sense. This latter kind of disparaging attitude toward the more-than-human world is rooted, Abrams argues, in a radical divorce of human beings from their natural environment. As much of modern civilization has gradually moved away from oral forms of knowledge transmission and more toward written modes of information and learning, Abrams suggests we have unwittingly ensconced ourselves in language that is almost exclusively interhuman. This shift in language and learning also marks a move away from a primarily bodily engagement with the natural world and toward a more exclusively cognitive and linguistic relation to human cultural products and relations. Abram argues that the consequence of these transformations is a widespread decline in the human ability to learn from the more-than-human world, to recognize our deep dependence on it, and to respect its richness and spontaneity.

Rather than recommending a simple return to oral literacy to solve these issues, Abram proposes developing new ways to employ written language (and related forms of knowledge and mediation) such that animals and the entire natural world are re-engaged in view of their wildness and difference. On this approach, animals and the natural world cannot be seen as something nonhuman, as if they lacked something human beings have. Rather, the natural world is here encountered as being "more-than-human," as something that *adds to* and *enriches* human relations and knowledge; thus, the phrase "more-than-human" carries the sense that animals and the natural world precede (historically) and exceed (epistemologically) human beings. Abram ultimately calls for ways of thinking, writing, and living that acknowledge this fact and that recall us to our "genesis in the interplay and tension between our animal body and the animate earth" (Abram 2005: 174). In his recent work, Abram has referred to this regenerative project as the "**becoming-animal**" of human beings (Abram 2010). With the concept of becoming-animal, Abram proposes a notion of human embodiment that is deeply and fundamentally animal in nature. To be an embodied human being is to inhabit a site that is shared between human and more-than-human beings, a site that returns human beings to a zone in which alternative, embodied, affective relations with animals and the more-than-human world become possible.

Abram's concept of the more-than-human world was initially taken up by ecofeminist Val Plumwood (2002) and several other prominent theorists. It has since played an important role for researchers in a variety of disciplinary contexts, including animal studies (Asdal, Druglitrö, and Hinchliffe 2017), environmental humanities (Bastian et al. 2016; Cianchi, 2015; Wright 2017), Indigenous studies (Johnson and Larsen 2017), sociology (Pyyhtinen 2016), and urban studies (Maller 2018). In this emerging body of research, the phrase more-than-human is generally used to indicate research programs that are dedicated to reconfiguring relationships with animals and nature in a more affirmative, ethical manner.

ONTOLOGY/ONTOLOGICAL

Ontology is a field of philosophy and theory that examines the most basic features of reality. There are several approaches to ontology, with some frameworks emphasizing relations, connections, processes, and assemblages as the most basic components of reality and others maintaining that individuals, substances, persisting objects, and the like are the most fundamental units.

In this book, the terms ontology and ontological are typically used in regard to the **human/animal distinction**. To inquire about the ontological status of human beings and animals is to ask such questions as: Are human beings and animals radically and essentially different, as dominant accounts argue? Do human beings and animals constitute a single, large set, as biological findings seem to suggest? Are the differences and relationships between and among human beings and animals so complex that more refined terms are needed to describe these realities, as many animal studies theorists maintain? In relation to these questions, we might also wonder: if we adopt a specific ontological framework, what are the implications of that framework for **ethics** and other value-laden interactions between human beings and animals? At present, there is no single, unified ontological account of human and animal existence that dominates in animal studies; instead, a number of competing frameworks and alternative ontological approaches are in dialogue with one another.

PETS

The term "pet" encompasses an enormous variety of human–animal relationships; hence, it is challenging to provide a simple definition that covers all the ways the term is used. Etymologists note that the term originally referred to spoiled children and then was extended to animals who were similarly indulged. This sense of the term is consonant with the way the word is employed in common parlance, where it is typically used to designate domesticated animals who live in the homes of their human companions and are treated with special affection. It can, however, also extend far beyond that usage to include animals who are caged or chained and kept outside the home and who are offered no special care. Among animal studies theorists, the term pet is a controversial one, as it carries connotations of human dominance over animals and reinforces the notion that pet animals are the property of human owners. Thus, animal studies theorists will often refer to pets as *companion animals*, a term that suggests human and animal relationships should ideally be conceived of more on the model of friendship than ownership. (In this entry, I will use both companion animal and pet depending on the context.)

Pets (and domesticated animals more generally) have played a significant role in the development of human history and civilization. Historian James Serpell notes that the domestication of animals for purposes of human companionship dates to more than 10,000 years ago and that evidence of such relationships can be found in cultures across the globe (Serpell 1986). Most scholars believe that dogs were the first animals to be domesticated and that they played the role of both companions and helpers (in the latter role, as protectors and as assistants in **hunting** and other similar tasks).

While pet-keeping has remained a consistent and fairly common practice throughout most of recorded human history, in the past century the presence of pets in human homes has skyrocketed. The most recent research from the American Veterinary Medical Association reports that over 50% of households in the United States have a pet, with dogs and cats being the most common pet animals (AVMA 2018). They also report a sharp rise over the past decade in the keeping of so-called specialty or exotic pets, such as fish and snakes. Similar trends are found in several advanced industrialized countries, with pet ownership being particularly high in Argentina, Mexico, and Brazil (GFK 2016). Accompanying this growth in pet ownership is an ever-increasing commercial sector dedicated to feeding and providing services and accessories for pets. In the United States alone, over $70 billion dollars were spent on pets in 2018 (APPA 2019).

These global trends indicate that pets have now become one of the key axes along which human–animal bonds are formed; as a result, many of our intellectual conceptions of who animals are and the affective relations we establish with animals occur within the context of pet keeping and ownership. However, as noted at the beginning of this entry, such relations of ownership and control pose a number of ethical problems. Ought we think of ourselves as owners of animals? What happens when economic forces come increasingly to drive the pet industry and when emotional and affective relationships with pet animals are monetized? What sort of social and normative standing do non-pet animals have in our lives and institutions?

Some critics of the practice of pet keeping and the pet industry argue that these institutions should be entirely abolished. In view of unethical breeding practices, widespread abuse and neglect of pet animals, and the commodification of human–animal relationships more generally, activists who oppose pet ownership maintain that animals should never be seen as things to be owned by human persons. For these critics, pet ownership is akin to chattel slavery, reducing a living individual to a mere piece of property to be used and disposed of as the owner sees fit (Francione and Charlton 2016). In response, defenders of the institution argue that pet keeping has many mutually beneficial aspects and that animals have benefitted in various ways from domestication (Budiansky 1992). If pursued thoughtfully and ethically, pet keeping can, they suggest, lead to human flourishing (consider, for example, the many documented health and psychological benefits humans receive by being in regular contact with animals [Fine 2019]) as well as animal flourishing (think, for example, of a well-trained animal joyfully performing an activity while in peak athletic form [Haraway 2008; Rudy 2011]). Consequently, these advocates argue we should reform human–pet relations in a more ethical direction rather than abolish them. Such reforms include the kinds of linguistic and behavioral changes noted at the beginning of this entry (namely, developing companionships and friendships with animals and moving away from an ownership paradigm) as well as contesting the economically exploitative aspects of the bourgeoning pet industry.

See also: companion species; entertainment

Further reading: Beck and Katcher 1996; Grier 2006; Overall 2017; Tuan 1984

PLAY

Play is one of the most fascinating and most perplexing features of the animal world. Biologists and ethologists who study the phenomenon have had difficulty arriving at a definition that covers all of its forms. The most influential definition is provided by ethologist Gordon Burghardt, who suggests play differs from other activities in: being irreducible to survival functions; being intrinsically rewarding; involving exaggerated or modified forms of normal activities (for example, play-fighting versus actual fighting); being repeated for prolonged periods of time; and occurring under relaxed conditions (Burghardt 2005). Less technically, biologist Nathan Lents suggests that while animal play is difficult to *define* it is not hard to *identify*: most of the time, "you know it when you see it" (Lents 2016: 21; cf. Burghardt 2005: 51).

What makes animal play particularly intriguing for scientists is the fact that it does not fit easily into adaptationist paradigms in biology (that is, frameworks that try to reduce behaviors and traits to having an exclusive role in terms of the survival and reproductive fitness of a given species). From a standard evolutionary perspective, play would seem to be a waste of energy that could be better spent doing more essential things, such as getting food or resting. Play can also seem to be a risky venture, in that it might distract an individual or group from attending to approaching predators or other such threats. Despite these paradoxes, play is commonly found among mammals and in the bulk of bird species; it has also been documented in certain reptiles, fish, frogs, and even invertebrates (Burghardt 2015; Zylinski 2015).

Of course, animal play is of interest not just to scientists—it fascinates the general public as well, as made evident by the enormous popularity of films and social media that depict animal play. With the current widespread availability of digital cameras and phones, instances of animal play that were once considered merely anecdotal have now been captured on film and circulate online among millions of individuals. Viewers are captivated by watching crows use various devices to "snowboard" on icy roofs, ducklings make repeated passes on play slides, and predator and prey animals (who would typically be engaged in life-and-death struggles) engaging in joyful games together.

Given the widespread existence of play among nonhuman animals (and human beings as well), biologists and ethologists have sought to develop explanations of its importance and function. No single, unified explanation of the function of play has emerged, but it is clear that play does have an important place in the development and maturation of individual animals and the cohesion of kin and social groups. In a survey of recent research on the topic, Lents suggests that play fulfills several essential roles for animals, from learning one's social rank and the bounds of appropriate behavior to forming social bonds and developing key cognitive and motor skills (Lents 2016). In his own summary of the findings on animal play, ethologist Marc Bekoff (who is also among the leading researchers on this topic) argues that animals engage in play because it allows them the opportunity to exercise, develop motor skills, and learn a sense of basic fairness and justice (Bekoff 2007).

Most animal play researchers insist, though, that these biological and physiological explanations do not fully capture *why* animals play from the animals' own point of view. Animals—human beings included—play simply because it is *fun*. Human beings and animals are complex creatures, and studying play behavior demonstrates that not everything we do can be fully explained in the objective language of biology. Some behaviors like play seem to go beyond direct evolutionary ends, and the joy that accompanies play seems to make the behavior sufficient unto itself.

Inspired by the joyful, social affects involved in animal play, animal studies theorists have looked to it as a way of rethinking the **ontology** of human–animal existence and as an ethico-political ideal for human–animal relations. With regard to ontology, theorist Brian Massumi (2014) has argued that play offers a provocative way of thinking about the overlapping modes of human and animal existence. For Massumi, play is an interesting phenomenon inasmuch as individuals do not so much "have" it as they are caught up in it; it is an activity in which both humans and animals participate and during which various affects and events flow through them. Play thus constitutes what Massumi describes as a zone of *mutual inclusion*, wherein human and animal cannot be clearly differentiated. Play also represents an interesting alternative site for rethinking ethical relations with animals. **Ethics** is most often understood as a set of practices grounded in the rather grave values of respect for and noninterference with others, or in compassion and **empathy** for the suffering of others. Rarely, though, do we think of play and joyful encounters as ethical ideals. Some animal studies theorists, such as Deborah Slicer, suggest that rather than focusing exclusively and critically on animals as victims of injustice, we should also consider more affirmative means of allowing for and co-creating more joy and play with and among animals (Slicer 2015).

Further reading: Bateson and Martin 2013; Bekoff and Byers 1998; Fagen 1981

POSTHUMANISM

Posthumanism names dramatic changes in our contemporary understanding of human beings and the nature of human relations with the nonhuman world. Proponents of posthumanism argue that classical conceptions of human nature—especially those based on human exceptionalism—are increasingly being called into question by advances in technology and science, as well as by transformations in ecological conditions and human–animal interactions. In line with the "post" in posthumanism, these proponents hope to inaugurate a new and different vision of what it means to be human, a vision grounded in the insight that human beings are becoming ever more indiscernible from the nonhuman world, both in its natural domains (plants, insects, animals, ecosystems, micro-organisms, and so on) and its humanly constructed, artifactual domains (technologies, media, practices, institutions, and so on). Posthumanism is of significant interest to practitioners of animal studies, as it places special emphasis on the need to rethink both the **human–animal distinction** and **anthropocentrism**.

The primary critical target of posthumanist theorists is the dominant image of human nature inherited from the classical Western tradition. In this tradition, human beings are typically understood to be fundamentally distinct from all other natural species as well as from human-made technologies and machines. What makes human beings exceptional on this account is their purportedly unique capacity for **language**, rationality, and autonomy. These distinguishing capacities, it is argued, also allow human beings to transcend their natural and material situations on earth.

At various points throughout modernity, this vision of the human has faced serious objections, with philosophers such as Julien Offray de La Mettrie (1996) arguing that human beings are essentially no different from machines, and scientists such as Charles Darwin (1981) proposing that human beings are fully natural, biological organisms (see **evolutionary theory**). In postmodernity, these challenges to human exceptionalism have accelerated. With regard to the human–machine distinction, technological advances have blurred this line such that what separates human from machine is no longer self-evident. The growth of artificial intelligence, robotics, medical technologies, gene modification, life-extension interventions, and other related technological advances undercut traditional ideas of where human nature ends and where machines begin. Today, increasing numbers of human beings are immersed in technologies and altered social environments from birth, rendering such individuals something closer to cyborgs than to intact members of a distinct and immutable human species.

One trend within posthumanism is *transhumanism*, which aims to extend and expand such technological modifications in view of perfecting human existence. Transhumanists argue that, with sufficient time and research, we will be able to put an end to many of the conditions that limit the lives and projects of finite human beings, such as an untimely death, useless suffering, debilitating injuries, and problems caused by genetically based diseases and disabilities (see Kurzweil 2005 for a popular version of this transhumanist vision). Although some transhumanists (for example, Pearce 1995) hope to use such technological advances to improve and perfect the condition of all sentient entities (both human and animal), animal studies theorists have been generally critical of the transhumanist project (Hauskeller 2017). Cary Wolfe (2010) has argued, for example, that transhumanism is not genuinely posthumanist inasmuch as it tends to reinforce many of the more problematic aspects of classical humanism. For Wolfe, transhumanism remains humanist especially in its desire to overcome our earthly, material, animal condition and to remove the fetters of embodied existence.

The form of posthumanism that has tended to be of most relevance to animal studies leaves behind this focus on perfecting human beings and looks instead to the technological and animal-biological disruption of traditional views of human nature as an opportunity to develop new ways of thinking and living. If it is granted that human beings are constituted from the ground up by a variety of relations (animal, biological, ecological, technological, and so on), then it would seem to follow that our worldview must shift accordingly and take a "nonhuman turn" (Grusin 2015). According to this "critical posthumanist" approach (Nayar 2014), the goal is to see reality as composed primarily of relations and interdependent sociality within which human beings find the conditions for their existence. A corresponding shift would also need to unfold in the ethical realm, with the primary aim being a reconsideration of what kinds of ethical responsibilities human beings might have to nonhuman others as well as relations that have previously been dismissed as different from and lower than the human (MacCormack 2012).

If this kind of critical posthumanist approach is pursued in the context of animals, Wolfe (2010) suggests that finitude, vulnerability, and mortality will not be seen as barriers to be overcome but rather as material, embodied realities that bind us with our fellow animals. Similar to the notion of **becoming-animal**, posthumanist thinking of this variety generally sets aside the project of trying to figure out what makes human beings unique and looks instead toward experimenting with different, more ethical relations with animals and the rest of the **more-than-human** world (in both the natural and artifactual realms). Such experiments with other ways of thinking and seeing are explicitly aimed at accelerating and deepening our posthumanist condition, seeking to complicate and render more complex our vision of human existence and the relations that make it possible (Braidotti 2013).

This variety of critical posthumanism has been criticized in fundamental, but charitable, ways from the perspective of critical **race** studies and from within animal studies as well. From the first perspective, Alexander Weheliye (2014) argues that posthumanism, in its rush to establish a posthumanist age, fails to consider other understandings of what it means to be human that derive from alternative, minor traditions. For instance, anti-colonial figures such as Frantz Fanon (1967b) and Sylvia Wynter (Scott and Wynter 2000) argue that classical humanism can best be overcome not by simply getting rid of the concept of the human but by rehabilitating humanism and giving it new meanings. In particular, they suggest that an alternative humanism constructed by those who have been most harmed by traditional humanism (the colonized, the marginalized, and so on) will provide a more just and less exclusionary vision of our common humanity. Critical posthumanists might wonder, in response, whether this project of rehabilitating humanism holds any promise for rethinking our **ontological** and ethical relations with animals and the more-than-human world.

From the second perspective, Donna Haraway (2016b) suggests that posthumanists remain too secure in their knowledge of what the human is and will become. For Haraway, "the human" is essentially a fabricated category, a fiction that tries to delink human beings from the relations that make them possible; in this sense, we have never been and never will be genuinely and properly human. What we call "human beings" are, for Haraway, better understood as "humus, not Homo, not anthropos" (Haraway 2016b: 55). In other words, instead of portraying human beings as a distinct species of beings separate from the earth who subsequently enter into relations, we should understand them as being entirely byproducts of the earth itself, as wholly animal-material creatures who belong to and are made of soil, humus. This alternative vision leads Haraway to describe human beings in the following terms: "we are compost, not posthuman" (Haraway 2016b: 55).

Further reading: Haraway and Wolfe 2016; More and Vita-More 2013; Pilsch 2017; Weinstein and Colebrook 2017

QUEER

In this entry, I follow contemporary queer theorists in using the term queer to refer to the multiplicity of sexual and gender practices, identities, and relations that depart from dominant norms (Halperin 1995). A frequent objection raised against individuals who manifest queer behaviors or identities is that their actions are "unnatural," which is to say, they run against the grain of supposedly universal biological sexual structures found throughout the human and animal worlds. The universal structures assumed operative here are those characteristic of heterosexual reproduction. The general idea is that across the biological world, sexual reproduction occurs exclusively between males and females who tend, by and large, to enact traditional kin roles (chiefly in relation to raising offspring). Sometimes this view of the purportedly universal nature of heterosexual reproduction moves beyond attempting to *de*scribe reality and seeks to *pre*scribe how human beings ought to behave. In such instances, heterosexual reproduction and its associated institutions become a normative ideal—hence the use among queer theorists of the term *heteronormativity* to describe this stance.

Biologists and ethologists have rigorously demonstrated, however, that this traditional picture of the supposed biological facts of human and animal heterosexual reproduction is fundamentally flawed. As scientists Bruce Bagemihl (1999) and Joan Roughgarden (2004) have persuasively shown, animal sexuality and kinship relations admit of enormous and astonishing variety. Whether it is a matter of sex characteristics, sexual behavior, gender, or parenting roles, the animal world complicates at every turn the notion that heterosexual reproduction and behavior is universal. From a macaque monkey mating with a sika deer, to fish species who change sex and mating roles within a single lifetime, to male and female parents who reverse parenting roles, animals exhibit sufficient examples of queer behavior to fill a multi-volume encyclopedia. Bagemihl's book *Biological Exuberance: Animal Homosexuality and Natural Diversity* (1999), which is perhaps the most exhaustive work on the topic to date, surveys the queer behavior of nearly 200 mammal and bird species—and this analysis does not even scratch the surface of the phenomenon (see Hird 2006 for a helpful survey of this material).

For those who endorse and defend traditional heteronormativity, one of the most disconcerting aspects of human queer relationships and ways of life are their often explicit refusal of the importance of forming a traditional family and having children. As Lee Edelman (2004) has argued, much of the dominant heteronormative culture is oriented around an ideal of *reproductive futurism* in which the primary value and meaning of social life is thought to reside in protecting the future for innocent children of privileged social classes (Edelman refers to these ideal children as "the Child" in order to mark their particular rather than universal character). In exhibiting their indifference to such familial and reproductive mandates, queer people are sometimes seen as being at odds in a fundamental way with the most important aims of civilized life. From a heteronormative perspective, they represent a dark and destructive set of sexual passions that are at bottom anti-Child and anti-civilization.

Animal studies theorists have found this queer analysis of reproductive futurism of interest inasmuch as being vegan and caring deeply about the fate of the nonhuman world is sometimes accused of sharing the same anti-Child and anti-civilization attitude. For example, to identify and live as a vegan is to challenge to some extent the **carnophallogocentrism** and **anthropocentrism** of the social order; for not only is veganism often associated with failing to achieve a fully masculine subject position, but caring about the well-being of animals for their own sakes involves turning away from an exclusive focus on the well-being of the traditional, privileged family and the Child (Dell'Aversano 2010; Simonsen 2012). In this sense, pro-animal advocacy joins queer theory and activism in challenging the dominant vision of what it means to live a worthwhile life and die a death worthy of grieving. While a concern for the flourishing of animals and a rejection of reproductive futurism among animal advocates is sometimes linked with anti-natalism (Benatar 2008), it is equally compatible with the notion that human beings should endorse a broader conception of life and reproduction and seek to "make kin [with the more-than-human world] not babies!" (Haraway 2016b; see also Clarke and Haraway 2018).

In relation to the queering of *gender* identity, Eva Hayward and Jami Weinstein (2015) argue that the category of trans[*] (written with an asterisk to underscore the prefixial and transformative nature of transgender identities) might help us to rethink what is at stake in animal studies and related post-anthropocentric discourses. They suggest that trans[*] discourse and concepts provide a way of thinking about human and animal gender identities as emerging from a series of dynamic relations rather than being indicative of some stable, underlying essence. For Hayward and Weinstein, gender identities are best understood as being inherently unstable and structurally susceptible to being unraveled by forces and relations that cut though and across them in various ways, thereby exposing individuals to unanticipated possibilities and potentials. Although trans* thinking is often thought to be limited to human gender categories, Hayward and Weinstein, along with other theorists like Mel Y. Chen (2012), maintain that the trans* dynamism at work in human registers crosses the human/animal divide and denotes more broadly the queer, messy, and unpredictable nature of life and matter as such.

To acknowledge that human and animal passions have inherently queer and trans* properties raises in turn the complex question of what happens when such passions cross species lines. There are, of course, many examples of interspecies sex, or sexual activities and relationships that occur between animals of different species (Gröning and Hochkirch 2008); similarly, sexual relationships of many sorts occur between human beings and animals (Beetz 2008). Despite widespread taboos against the practice, bestiality has in fact been practiced throughout human history and in a number of cultures (Miletski 2005). And while the dominant culture and the pro-animal movement are generally at odds when it comes to what constitutes proper treatment of animals, on the issue of bestiality they tend to share the same general revulsion toward the practice. It is generally assumed by both groups that human beings having sexual relations with animals involves taking advantage of vulnerable animals and violating an animal's autonomy (although, as many pro-animal advocates would point out, several other violent human–animal interactions such as eating meat involve the same ethical violations).

But this categorical rejection of bestiality and zoophilia has been challenged from various quarters, especially by those who call themselves zoosexuals (or "zoos," for short). For zoosexuals, sexual relationships with animals can and sometimes do develop out of a genuine love and respect for animals. Such relationships are not paternalistic, zoosexuals insist, because they stem from a perspective and disposition that grants animals a strong sense of **agency** and personhood. In this sense, zoosexuals argue that their sexual practices are organic outgrowths of mutual human–animal affection and are consistent with the deepest principles animating animal **ethics** (Cassidy 2009).

This controversial position has found partial support in the writings of Peter Singer, who is one of the most influential philosophers in the field of animal ethics (see the entry on **animal liberation** for more on his work). In an infamous article entitled "Heavy Petting,"

Singer (2001b) suggests that not all human–animal sexual relationships can be considered harmful or unethical. In fact, as Singer notes, sometimes animals themselves initiate sexual relationships with human beings. While Singer does not wish explicitly to endorse bestiality or zoosexuality, he does make the point that we should not be disgusted at the prospect of human–animal sexual relationships. Human beings are fully animal for Singer, and there is no natural, biological reason that sexual passions must respect the species barrier. While Singer's reflections on bestiality are broadly consistent with a utilitarian ethic (one of the most influential normative theories in philosophy), his thoughts on this topic have been roundly criticized by many of his fellow animal ethicists and pro-animal advocates (see Levy 2003 for a fuller discussion). These ongoing debates over bestiality and the implications of queer and trans[*] approaches to animal passions are clearly among the most contentious and charged issues in the field, but they are important ones for coming to grips with some of the unanticipated implications that derive from attributing agency to animals and blurring sharp **ontological** distinctions between human beings and animals.

See also: human/animal distinction; intersectionality

Further reading: Alaimo 2010; Dekkers 1994; Giffney and Hird 2008

RACE

The relationship between animals and race is one of the thorniest issues in animal studies. The discussion has become even more fraught with tension in recent years due to high-profile animal activist campaigns and theoretical work in animal studies that posit fundamental links between racism and **speciesism**. In 2005, the pro-animal organization People for the Ethical Treatment of Animals (PETA) launched a much-publicized campaign called "Are Animals the New Slaves?" and followed that campaign with several others that sought to link slavery, the Holocaust, and other forms of violence against racial and ethnic minorities with the mistreatment of animals. In these campaigns, PETA used extraordinarily graphic depictions of human slavery, brutality, and **captivity** alongside similar images of the mistreatment of animals to highlight parallels in the techniques and ideologies of human and animal oppression.

Such campaigns have direct counterparts in the animal ethics literature, where the mistreatment of animals is often explained and criticized in relation to unjust race-based forms of discrimination. From classical thinkers such as Jeremy Bentham and Henry Salt to contemporary pro-animal advocates like Peter Singer and Gary Francione, the violent capture and treatment of animals has been recurrently compared with the enslavement of Black human beings, with the explicit understanding that both injustices should be rejected for the same basic reasons (namely, that discrimination of this sort is a violation of the interests of both human and animal individuals). In 1988, pro-animal author Marjorie Spiegel published *The Dreaded Comparison: Human and Animal Slavery*, which made extensive visual and theoretical connections between the enslavement of Blacks and the degraded status of animals. This book summarized the standard animal ethics approach to thinking about race and animals and has frequently been used as a template for subsequent theoretical work on the topic and for activist campaigns like PETA's. (It should be noted that Spiegel objected to this use of her work by PETA in particular and sued them.)

By and large, these analogical arguments have not been well-received within movements for racial and social justice. Even among critical race theorists who take animal issues seriously, this particular way of aligning the two movements has been subjected to severe scrutiny. One of the common refrains issued by activists and scholars of color in this regard is that racism and slavery are often used in an instrumental manner by pro-animal authors and activists who make analogical arguments. Although pro-animal advocates insist in response that their discourse is equally committed to anti-speciesism *and* anti-racism, critics charge that there is little evidence of sustained engagement with anti-racist struggles by mainstream animal advocates. At the same time, critics note that shocking and painful images depicting slavery and racist violence are often used by animal activists without considering the psychological impact such media might have on people of color. The general sense, then, is that the campaigns and arguments by animal advocates that align racism with speciesism are not genuinely interested in the mistreatment of people of color but are primarily piggybacking on racial justice movements in order to make the pro-animal cause more visible (see Boisseron 2018 and Johnson 2017 for fuller discussion of the foregoing themes).

Another critique of the facile analogizing of racism and speciesism concerns how racial injustice and **dehumanization** are understood by some animal advocates. In the animal ethics literature, speciesism is often characterized as the last remaining form of prejudice, with the implicit assumption being that **equal consideration** and full humanity have effectively been granted by the established order to racial minorities and other marginalized peoples. But as Alexander Weheliye argues, it is a profound mistake to assume that all human beings "have been granted equal access to western humanity" (2014: 10). To the contrary, as Weheliye and other critics note, many people of color live in social conditions where their humanity is constantly called into question and where they are subject to being harmed and killed with impunity.

Despite these problems with analogizing racism and speciesism, there remain compelling reasons for thinking about race and animals in conjunction. Informed by the latest developments in critical race theory and animal studies, a new generation of theorists and activists have sought to recast and rethink this relationship. Their approach focuses not on the privilege of the human species as a whole but on the privilege of "Man," where Man refers to traditional notions of humanity that are based on masculine ideals, norms of whiteness, and a humanity that is dominant over and separate from animal existence (Polish 2016; Ko and Ko 2017). According to these theorists, people of color and animals both serve as Man's "Other," that is, as beings who are marginalized in order to guard the propriety of humanity. Inasmuch as animals and people of color occupy this shared space of violent exclusion and differentiation from the order of Man, they often suffer interconnected forms of oppression and violence. Implied in this recasting of the relation between animals and race is a move away from a critique of speciesism (which assumes that all members of the human species have full and equal standing and that animals do not) toward a critique of **anthropocentrism**. The latter concept suggests that the dominant culture and its main institutions revolve around the interests of a small, privileged number of human beings (White, able-bodied, male, meat-eating, and so on) at the expense of the vast majority of humanity (especially the dispossessed and marginalized) and the more-than-human world (with animals being a prime but not sole figure of this excluded class).

In order to articulate the complex relations and intersections among Man's Others, theorist Claire Jean Kim (2015) calls for a *multi-optic* framework, in which multiple concerns, oppressions, and injustices are simultaneously held in view. For Kim, it is essential to move away from critical theories that are based on simplistic dualisms (for example, human/animal, human/nature, white/non-white) and develop more complex taxonomies of power. In regard to animals, for example, the point is not to critique and contest a simple human/ animal binary where all human beings have the same ontological and ethical status over and against animals (as the notion of speciesism implies); rather, according to Kim's analysis, we need to appreciate that the human is already internally fractured along racial and other lines, with various divisions that associate certain human beings with animals while also humanizing certain animals. Further, these fractures and divisions are not always consistently applied and uniformly valued in different contexts. Such complex power relations, then, can only be grasped with a correspondingly nuanced optic that emerges through careful analysis of the power dynamics at hand in a given situation. Kim argues that this multi-optic approach is most effective when it is coupled with an *ethics of avowal*, by which she means a willingness on the part of radical movements for social and more-than-human justice to consider the ways in which joint forms of oppression and strategies for resistance overlap.

See also: anthropological machine; biopolitics; intersectionality

Further reading: Boisseron 2018; Cordeiro-Rodrigues and Mitchell 2017; Harper 2010; Harper 2011; Jackson 2013; Peterson 2013; Socha 2013

REWILDING

The term rewilding was first used in the 1990s to describe strategies employed by environmental activists associated with *Earth First!* Beyond the goals of merely slowing **extinction** and limiting ecologically damaging activities, *Earth First!* activists proactively sought to restore and return the earth to a previous, wilder state (Foote 1990). Since its initial coinage, the concept has been enthusiastically adopted and significantly refined by environmental activists as well as scientists, journalists, and organizations of many stripes. Contemporary rewilding projects are of considerable interest to animal studies scholars, as they tend to place animals—especially megafauna, or large animals—at the heart of their reparative projects (Donlan et al. 2006). As such, rewilding stands out from traditional ecological restoration projects, which focus mainly on restoring native species of plants and nutrient cycles. Further, rewilding is much more ambitious in scope than traditional restoration ecology, taking as its goal the restoration not just of particular, local ecosystems but the rewilding of entire bioregions (Keulartz 2018).

The overarching ethical and aesthetic vision behind rewilding is the reconstitution of natural biodiversity and ecosystem health on a large scale, with the ultimate aim of providing more space and autonomy for animals and other nonhuman species. The means proposed to achieve these aims differ depending on the conception of rewilding under discussion. The best-known version of rewilding employs top-level predator species to create "trophic cascades" (that is, downward ripple effects in the food and energy systems within a given territory) that eliminate excess numbers of species at the next trophic level down, and so on down the chain. By restoring balance to these trophic relations, biodiversity of many kinds has the opportunity to flourish. This approach is commonly referred to as the *Three Cs* model, as it protects Core wilderness areas that are essential for healthy animal habitats, establishes Corridors to connect those core wilderness areas and allow animals to migrate between them, and employs Carnivores to stabilize trophic relations (Soulé and Noss 1998). Another common form of rewilding focuses on what is called *naturalistic grazing*, in which large herbivores are considered the main drivers of ecosystem health. In this approach, energy and food availability at lower trophic levels (for example, vegetation) constrain and determine what constitutes a wild and relatively balanced area. By allowing large herbivores to graze freely, the trophic chain becomes balanced and stabilized at lower and higher trophic levels and a sustainable number of species and individuals are gradually restored in a given region (Hodder and Bullock 2009).

These rather different approaches to rewilding point toward one of the deeper difficulties for rewilding in general, namely, how to determine the historical benchmark for what constitutes wildness. For the "Three Cs" model, which tends to dominate in North American contexts, the historical benchmark is often taken to be the late Pleistocene era, before widespread ecological changes and megafauna extinction were brought about by human beings in the Holocene. For the naturalistic grazing model, the mid-Holocene is the chosen benchmark, as it marks the era just prior to the widespread emergence of agrarian ways of life that radically changed wild landscapes. Whatever benchmark is chosen, there is no uncontroversial historical point at which we can find something like pure wilderness that is uninfluenced by human contact once *Homo sapiens* arrives on the evolutionary scene. And even if such a point were to be agreed upon by rewilders, it is evident that trying to reinstitute such a state of wilderness today would be extraordinarily difficult given the size of the current human population and its profound ecological influence (Toledo, Agudelo, and Bentley 2011).

Rewilding has been challenged on additional grounds, with critics arguing that these projects: lack public support for establishing large-scale wilderness corridors; offer insufficient scientific evidence to support their purported outcomes; contain an inherent bias against Indigenous peoples who have lived with animal species for millennia and often without causing serious biodiversity loss; and create intractable conflicts between wildlife and domesticated animals and people (see Lorimer et al. 2015 for further discussion). Critics also charge that rewilding discourse is plagued by a sharp and untenable division between nature and culture, with **wildlife** and wilderness entirely on the nature side of the binary and all human activity on the culture side. In response to this sharp division, theorists argue that in an era in which the effects of human activity have penetrated ever deeper across the globe, there is little point in talking about a natural world or wilderness devoid of human culture (Jørgensen 2015; see Prior and Ward 2016 for a contrasting view). Developing this point further, there are concerns even among those who support the ideals of rewilding that the aims of such projects are likely to be foiled by rapidly accelerating human-caused climate change and ecological degradation (Corlett 2016).

Refining the notion of rewilding in view of these criticisms, ethologist and long-time animal advocate Marc Bekoff suggests that the promise of rewilding lies not in turning back the clock to some arbitrary historical benchmark or in creating a world in which wilderness remains utterly free of human influence. Instead, he argues that rewilding should be understood as a forward-looking ideal, an attempt to "make room for much more diverse, healthy, and sustainable ecosystems that are as natural as they can be given [human] omnipresence" (Bekoff 2014: 10). In addition, Bekoff believes that rewilding should not be limited to the task of creating space for the repair and restoration of the natural world alone; human beings, too, need to undergo this kind of radical transformation. Bekoff refers to such changes in human beings as "rewilding our hearts," by which he means undergoing a personal transformation and change of heart such that we come to reconnect with and develop love, **empathy**, and a sense of friendship for our animal kin.

See also: Anthropocene; endangered species; ethology; extinction

Further reading: Foreman 2004; Jepson 2016; Monbiot 2014

ROADKILL

Roadkill is a colloquial term for animals who have been struck by automobiles and left in or near the roadway to die. Although roadkill is sometimes used in reference to domesticated animals and **pets**, the term is most commonly applied to **wildlife**. The word roadkill is relatively recent in origin, emerging in the mid-20th century when automobile driving and highway construction became more common and increasing numbers of automobile–animal collisions began to occur. There is no national or international agency that tracks the number of animals killed on roadways each year, but it has been estimated that some one million animals per day become roadkill in the United States alone (Lalo 1987; see Seiler and Helldin 2006 for a discussion of roadkill numbers in an international context). If this estimate is even somewhat accurate, roadkill might outstrip **experimentation** and **hunting** in terms of the sheer number of annual animal **deaths**.

Even though automobile driving clearly has a substantial negative impact on animal well-being, the topic has received surprisingly little attention in animal ethics and animal studies until recently. One of the scholars who has done groundbreaking work in this area, Dennis Soron, suggests the issue of roadkill has been neglected because it does not conform to the established view among pro-animal organizations of why animals are treated with violence or how such violence might be abated (Soron 2007). Mainstream approaches to **animal welfare and animal rights** have tended to see violence toward animals as the result of an individual moral failing that can be addressed through making changes in personal habits of consumption or supporting corrective legislation (for example, as an individual I can avoid contributing to the harms caused by **CAFO**s by refusing to purchase meat or supporting legislation aimed at reforming CAFOs). But with roadkill, the situation is more complex. Roadkill is not simply the result of an individual moral failing, and it cannot be solved by some individuals avoiding driving or making minor reforms to traffic laws. This is because current driving habits and practices emerge out of a larger system that has come to dominate contemporary societies, a system which sociologists Peter Freund and George Martin refer to as *hyperautomobility* (Freund and Martin 2007). Within this system, the rapid transmission of human bodies and economic goods becomes the primary goal, with the well-being of animals and the more-than-human world placed largely out of consideration. In hypermobile cultures, roadkill comes to be seen as collateral damage, a regrettable but ultimately unavoidable side-effect of dominant practices of mobility. Addressing the problem of roadkill within this context would thus require rethinking our entire way of life, both in terms of the value attributed to ever-increasing mobility and to the priority given to the capitalist economic practices that undergird and encourage increased mobility.

As animal studies theorists and other concerned academics and activists have begun to reflect more carefully on the problem of roadkill, a number of responses and lines of inquiry have emerged. At the level of mobility infrastructure, a sub-discipline of ecology called *road ecology* has dedicated itself in part to creating roadways that are more attentive to the welfare of wildlife. Road ecologists design various apparatuses to help wildlife safely cross roads (using tunnels and bridges, for example) or stay clear of roads altogether (with the help of fences and barriers, or redirecting roads away from animal habitats). They also help to build infrastructure that minimizes pollution and limits habitat fragmentation (van der Ree, Smith, and Grilo 2015).

Other thinkers who have given serious consideration to roadkill, such as environmental writer Barry Lopez, believe that it is necessary to take up a more mindful, even apologetic stance toward the animals killed on our roadways. In his short essay, *Apologia* (1998), Lopez takes the reader on a car trip across the United States and narrates his numerous encounters with roadkill. Lopez ritually removes the roadkill he encounters and wonders what else can be done to atone for the deaths that he and his fellow drivers cause. With similar concerns in mind, a number of artists have recently sought to call attention to roadkill through adorning dead animal bodies on the side of the road, creating roadkill photograph exhibits, and repurposing roadkilled animals for various art projects. The aim of much of this **art** is to recall the viewer to the materiality, singularity, and uncanny beauty of the animals that are left for dead on our roadways. By staging encounters with roadkill through art, **literature**, and other means, those who are concerned about the fate of roadkill hope to create the conditions for making roadkilled animals matter in society

(Desmond 2016). At the level of individual and collective transformations, activists are engaged in experiments with different mobility systems that cause less damage to animals and the ecology; they are also creating coalitions with local and national movements for traffic justice. In view of the fact that a hypermobile culture makes use of numerous forms of transportation—such as planes and trains—that also cause animals serious death and injury, historian Gary Kroll has suggested expanding the discussion beyond car-based roadkill. He recommends the term "snarge" (a word initially coined to refer to the remains of a bird that has been struck by a plane) be employed to refer to the wide range of collisions that occur between various kinds of "fossil-fuel-based human mobility and solar-based animal mobility" (Kroll 2018: 82).

Further reading: Clark 2017; Desmond 2016

SENTIENCE

Sentience refers to the capacity of an individual organism to sense and feel things and to have subjective experiences of those sensations and feelings. In debates in animal ethics, what is most often at issue with this term is whether animals are able consciously to experience pain and pleasure and have conscious preferences for one state of being over another. If it is the case that a given animal *is* sentient in this robust sense of the term, most animal ethicists argue that we have strong reasons for considering how our actions might affect that animal. Although animal ethicists make use of a number of different ethical frameworks, nearly all ethical approaches are united in the idea that causing unnecessary suffering to sentient beings is something to be avoided and that, conversely, causing suffering requires some form of justification. A handful of philosophers and scientists argue that animals are not technically sentient because they seem to lack subjective experiences of pain; however, most of the research in animal ethics and animal studies presumes that the bulk of the animals that are consumed, experimented on, and used for **entertainment, work**, companionship, and so on are strongly sentient (for a helpful introduction to discussions about the nature and existence of animal sentience, as well as a consideration of whether fish feel pain, see Braithwaite 2010). Therefore, animal ethicists argue that we ought to give serious consideration to curtailing or even abolishing practices that cause animals pain if we lack compelling reasons for continuing such practices.

A key question about sentience in animal studies is whether sentience should be the sole criterion in ethical matters. For figures like Peter Singer (2001a) and Gary Francione (1996), sentience is not only sufficient for being deserving of ethical consideration, it is necessary—which is to say, if a being is *not* sentient, then on their account it should have no ethical standing. This position, which limits ethical consideration to sentient beings alone, is sometimes referred to as *sentientism*, or *sentiocentrism*. It is often contrasted with ethical frameworks that have a broader scope of concern, such as biocentrism (which includes all of life within the ethical domain), ecocentrism (which makes ecosystems the central focus of ethics), and phytocentrism (which centers ethics on plants). These broader frameworks usually include animals within their scope of concern but do not limit ethical consideration to sentient animals alone.

See also: abolition; environmentalism; equal consideration; ethics; mind; welfare

Further reading: Carruthers 1992; Dawkins 1980; Varner 2001; Warren 1997

SHELTERS AND SANCTUARIES

Animal shelters and sanctuaries protect animals whose lives are in precarious conditions. Shelters are somewhat different from sanctuaries in that the former are usually short-term sites for holding animals whereas the latter serve most often as long-term or even permanent safe havens. Shelters take a variety of forms, from publicly funded shelters that accept nearly all abandoned animals to private shelters that take in only animals that can be readily adopted. They have their historical origins in welfare organizations that sought to shield abandoned urban animals from cruel treatment (for a helpful and fuller analysis of the history of shelters, see Irvine 2017). Contemporary shelters, however, typically focus their efforts not just on stopping the abuse of stray animals but on placing animals in loving and caring homes. To this end, some shelters engage in public outreach to encourage individuals and families to adopt abandoned animals as **pets** and also promote spaying and neutering to reduce the number of unwanted animals.

Many public shelters find themselves in the position of having to deal with animals who are not adoptable (either because they are deemed too dangerous to be adopted or because they are simply unwanted). This problem has led many shelters to euthanize such animals (some 1.5–2 million shelter animals per year are euthanized in the United States alone). Euthanizing unwanted shelter animals is controversial with most animal rights advocates, who believe that shelters should adopt a "no-kill" policy and find alternative ways of placing animals in homes (Winograd 2007). Defenders of euthanasia argue in response that there will always be unadoptable animals and that the most compassionate practice is to euthanize them. This split on the ethics of euthanasia for shelter animals reflects a schism among pro-animal activists that is marked in the distinction between **animal welfare and animal rights** (Arluke 2003). (For more on the recent, high-profile controversy over People for the Ethical Treatment of Animals' participation in euthanizing animals, see Newkirk 2019 and Winograd and Winograd 2017.)

Animal sanctuaries have become increasingly common in recent years as a means of providing living spaces for vulnerable animals who do not typically find homes in urban settings. Such animals might include those rescued from slaughterhouses or experimental laboratories as well as wild animals who have been injured or who have lost their native habitats. Some of the more prominent animal sanctuaries see their work as an important form of **activism** and an effective means of responding to widespread violence against animals. For example, Farm Sanctuary, founded by Gene Baur and Lorrie Datson, cares for farm animals that have been rescued from slaughterhouses and uses its facilities to promote interaction with farm animals and educate the public on the intellectual and emotional lives of the animals. Similarly, the VINE (Veganism Is the Next Evolution) sanctuary, founded by pattrice and Miriam Jones, provides care for a variety of rescued animals while promoting veganism and animal advocacy based on a politics of **intersectionality** and social justice.

Despite the unquestionably important services that shelters and sanctuaries provide for many animals, even the staunchest advocates of these institutions find themselves in complicated ethical positions with regard to the animals they house and protect. On the one hand, by promoting the adoption of animals, shelters tend uncritically to reinforce the notion that animals should be pets for human beings and properly belong in domesticated settings. Further, by committing themselves to reducing the number of unhoused animals through promoting spaying and neutering, shelters and their advocates are directly manipulating and interfering with the reproductive autonomy of animals in questionable ways (see the entry on **biopolitics** for a broader discussion of the concerns surrounding the control of animal life). On the other hand, sanctuaries run the risk of becoming not just long-term safe havens for animals but permanent sites of **captivity** where animals are held in fixed sites against their own will (but, presumably, for their own good); such practices can lead to paternalistic attitudes toward animals, reinforcing the idea that human beings should decide the living conditions of animals. Advocates for shelters and sanctuaries are often aware of and share such concerns but see these institutions as interim solutions on the path toward transforming societies in the direction of allowing for increased animal **agency**.

Further reading: DeMello 2017; Emmerman 2014; Reese 2018; Sandøe, Corr, and Palmer 2016

SPECIESISM

A term with a wide range of meanings and uses, speciesism was first used by Richard Ryder (1971) to describe an unjustified bias against animals based on the purported superiority of the human species. Peter Singer (2001a) subsequently popularized the term, defining it as a prejudice that grants ethical consideration solely to humans on the basis of species membership. The prejudice here lies in the arbitrary denial of fundamental ethical similarities among human beings and animals that cut across the species barrier.

The term's suffix, "-ism," is meant to recall other discriminatory "-ism"s such as racism and sexism. These prejudices are considered unjustifiable because they afford certain groups (in this case, white persons and men) higher ethical standing than their counterparts (persons of color and women) without persuasive reasons for doing so. (The underlying assumption is that if an individual or group is to be given lesser ethical standing, there must be good reasons for this reduced standing.) But if racism and sexism are undercut by the preponderance of similarities among human beings of different races and sexes, wouldn't the manifest *dis*similarities separating human beings from animals serve as grounds for *justifying* speciesism? Critics of speciesism respond that the sorts of differences between human beings and animals typically emphasized (such as higher-order consciousness or **language**) do not constitute grounds for excluding animals from ethical consideration. At the level of what is ethically relevant—subjectivity, **sentience**, relationships, and so on—human beings and animals are sufficiently similar so as to render species-based discrimination unjustifiable.

Since its initial formulation, the term has undergone a number of modifications, expansions, and challenges. Joan Dunayer (2013), for instance, argues that the kind of speciesism described by Ryder and Singer should be labeled "old speciesism." Such speciesism is easy to criticize and reject because it is based on outmoded claims about human–animal discontinuity that fly in the face of recent biological and empirical evidence. More problematic for Dunayer is the "new speciesism" characteristic of the work of some animal advocates who extend ethical consideration only to those animals who most resemble human beings (for example, higher-order primates, mammals, and so on), leaving other sentient animals outside the scope of consideration.

Working from within the perspective of sociological theory, David Nibert (2002) argues that defining speciesism as an individual prejudice (as many animal ethicists do) leads to a mistaken account of why animals are mistreated. Nibert suggests that speciesism is better understood as an *ideology*, that is, as a set of shared ideas and beliefs that function to justify a given socio-economic order. From this standpoint, speciesism is less a matter of individual behaviors and attitudes and more a matter of economic and structural injustice and is most effectively addressed and contested as such.

See also: animal–industrial complex; capitalism; equal consideration; ethics; race; feminism

TOTAL LIBERATION

A term employed by theorists and activists in anarchism and critical animal studies and popularized by philosopher Steven Best (2014), total liberation includes but goes beyond **animal liberation** to defend a broad liberation platform. In this approach, animal liberation is understood to be but one liberation movement among others, but it has a central role to play in highlighting the need for radical and progressive social movements to move beyond their exclusive focus on human politics and to consider the well-being of the nonhuman world. At the same time, total liberation challenges animal liberationists to extend their scope of concern to embrace environmental and ecological perspectives and to recognize their shared heritage with other radical movements for social change.

Total liberation is grounded in an anarchist framework that seeks to root out and contest all forms of domination. As opposed to traditional forms of anarchism, however, the total liberation approach understands domination and oppression to apply to both the human and nonhuman world and seeks justice for the nonhuman world as well. Total liberationists argue that achieving this broader conception of liberation involves developing a wide number of political alliances and coalitions. As such, total liberationists maintain that rights and liberation movements of many kinds (such as those associated with **feminism**, environmentalism, **disability** issues, class, **race**, animals, and so on) need to work jointly to challenge inequality and hierarchies and develop a more just and sustainable way of life. It is important to note that, according to total liberationists, the linkages between these liberation movements are not simply a matter of shared names or ideologies. What links the movements is ultimately a shared economic and institutional enemy—namely, the capitalist economic system and modern nation-states and militaries (which function respectively as **capitalism**'s political representatives and means of force). For total liberationists, these economic and political systems are at the heart of the violent and exploitative treatment of human beings, animals, and the earth and must be directly challenged and replaced with a non-exploitative alternative.

One of the distinctive aspects of the total liberation approach is its commitment to multiple kinds of **activism**, including (most controversially) militant direct action. Given the deeply violent and amoral nature of contemporary capitalism and state politics, total liberationists argues that tactics such as breaking into labs or factories to free animals and damaging property and research facilities are necessary for activism to be effective. Similarly, total liberationists take issue with forms of animal liberation and **abolitionism** that focus on incremental economic and political reforms. From the total liberation perspective, having individuals adopt vegan diets or vote for the abolition of certain violent animal practices are inadequate strategies, insofar as they leave in place the very economic, political, and legal systems that have given rise to anti-human, anti-nature, and anti-animal violence. Critics of the total liberation perspective (for example, Pellow 2014) argue that its attempt to determine what form liberation should take for all human, animal, and earthly beings and systems risks becoming its own form of imperialism, imposing a single and unified social, economic, and political perspective on the entire planet.

See also: environmentalism; intersectionality; politics

Further reading: Best and Nocella 2004; Best and Nocella 2006; Nocella, White, and Cudworth 2015

VEGANISM AND VEGETARIANISM

Veganism and vegetarianism are terms used primarily to refer to ways of eating that abstain from the consumption of animals. Vegetarianism is focused on avoiding animal flesh, whereas veganism goes further and avoids the consumption of flesh as well as animal byproducts (such as milk and eggs). Although veganism and vegetarianism can both be understood mainly as dietary choices, veganism in particular often includes a broader commitment to an alternative way of life (McKay 2018). This entry begins with a brief overview of vegetarianism and its variants before exploring veganism both as a way of eating and a way of life. It concludes with a discussion of some of the objections raised against veganism by animal studies scholars.

As just noted, vegetarianism in the most common sense of the term names a diet that avoids the consumption of **meat**; but vegetarianism is practiced with varying degrees and kinds of abstinence. *Lacto-ovo vegetarians* abstain from meat, poultry, and fish but consume eggs and dairy; *pesco-vegetarians* abstain from meat and poultry but consume fish, eggs, and dairy; and *flexitarians* eat a predominantly vegetarian diet but also occasionally consume meat. While many vegetarians cite an ethical rationale—namely, concern for animal welfare—for adopting their diet, vegans maintain that the production of animal byproducts should also be eliminated on these same grounds. Vegans thereby argue that the violent and unethical nature of modern industrialized animal agriculture is not limited to the production of meat. Battery hens (caged chickens that lay eggs), for instance, are kept in horribly cramped conditions, subjected to painful debeaking, and often suffer from painful physical ailments during their short lives; and dairy cows who live in **CAFO**s are forcibly impregnated in order to induce lactation, confined largely indoors, milked by machines, and sent off to slaughter as soon as the production of their milk supply becomes unprofitable for the farmer.

For many vegans, these ethical concerns about the violent treatment of animals provide sufficient reason for avoiding the consumption of animal flesh and byproducts altogether (hence, the often close relationship between veganism and **abolitionism**). But there are several additional reasons that some vegans cite for adopting their diet. In recent years, the causal role that animal agriculture plays in the acceleration of ecological degradation has come increasingly to light, leading many concerned individuals to adopt a predominantly vegan diet (see the entry on **climate change** for a fuller discussion of this issue). Others who are concerned about sexism, racism, social justice, and problems with global hunger have argued that a vegan diet is consistent with a more equitable way of life and sustainable distribution of food (Harper 2010; Robbins 2010; Adams 2015). Still others cite religious and spiritual reasons for avoiding the consumption of animals, maintaining that a diet that minimizes cruelty allows for a more compassionate disposition toward all forms of earthly life (Walters and Portmess 2001).

When Donald Watson coined the term "vegan" in 1944, he placed the dietary distinctive of abstaining from the consumption of all animal flesh and byproducts in the forefront of the term's meaning. For Watson, vegan eating was the logical outcome of the principles underlying ethical vegetarianism. He argued that being a vegan was the beginning and the end of being a *veg*(etari)*an* and that serious vegetarians should ultimately avoid consuming animal products of all sorts. But even at this early date, the term veganism was meant to indicate more than just a dietary distinctive. With his colleague, Elsie Shrigley, Watson co-founded The Vegan Society, which promoted veganism as a complete way of life dedicated to eliminating the exploitation of animals in all domains. The Vegan Society still promotes this broad ideal today, suggesting that veganism is a "philosophy and way of living which seeks to exclude—as far as is possible and practicable—all forms of exploitation of, and cruelty to, animals for food, clothing or any other purpose; ... by extension, [veganism] promotes the development and use of animal-free alternatives for the benefit of humans, animals and the environment" (The Vegan Society n.d.).

This emphasis on veganism as a practice aimed at ending the exploitation of animals *in toto* has, according to animal studies scholars Lori Gruen and Robert C. Jones (2015), given rise to two distinct types of veganism. On the one hand, there is the sort of veganism that argues for strict adherence to non-violent eating principles and that sees veganism as the "moral baseline" (Francione 2008) for any individual who claims to live and work in defense of animal justice. This sort of veganism, which Gruen and Jones call *Identity Veganism*, aims at achieving clean hands and a good conscience concerning the consumption of animals. For identity vegans, if one adheres to a strict vegan diet and uses "cruelty-free" cosmetics and other such consumer goods, one can maintain a consumerist lifestyle in good conscience, confident that one's purchases do not involve causing direct harm or death to animals. Another kind of veganism, which shares the same ideal of trying to end the exploitation of animals, sees violence toward animals as being a much deeper problem, one that demands thoroughgoing and long-term individual and collective transformations in response. This second kind of veganism, dubbed *Aspirational Veganism* by Gruen and Jones, treats veganism as an ideal that one continually strives after rather than achieves. Aspirational veganism places an emphasis on veganism being an ongoing process, a set of imperfect practices whereby one does as much as possible to minimize animal exploitation in one's life and in society more broadly. Here veganism becomes less of a moral baseline and more of a calling to an entirely different way of being in the world, where concern for animals, human beings, and the environment as a whole are continually negotiated and pursued in relation to other ideals of social justice and flourishing.

Animal studies scholars are not entirely unified in supporting veganism, in either the narrow identity or broader aspirational form (Warkentin 2012). Scholars such as Donna Haraway (2008) and Kathy Rudy (2011, 2012) question the premise that killing animals in order to consume them is necessarily unethical. They argue instead that problematic violence against animals stems from seeing animals as being *merely* or *solely* "killable," which is to say, with seeing animals as beings who can be killed with impunity (see V. Stanescu 2013 for a critical engagement with Rudy's work). In line with the sentiments of Haraway and Rudy, philosopher Dominique Lestel (2016) argues that vegan efforts aimed at removing animals entirely from the domain of edibility betrays a certain naïveté on the part of vegans and a willful disavowal of the fundamentally violent nature of embodied, earthly existence. Lestel goes so far as to suggest (with deliberate irony) that a respectful *carnivorous* diet actually demonstrates more solidarity with and love for animals than entirely abstaining from consuming them. From a slightly different angle, ecofeminist philosopher Val

Plumwood (1996) argues that fully acknowledging the edibility of animals need not lead to the adoption of a carnivorous diet. Instead, Plumwood suggests that if we appreciate the *shared* edibility of human beings and animals, then we might adopt ways of eating and living that are more respectful of animals, allowing them to exist as something more than mere meat to be consumed.

See also: activism; animal welfare and animal rights; companion species; feminism; intersectionality; meat; race

Further reading: Gillespie 2018; Wright 2015

WAR

When animal studies scholars and activists reflect on the nature and scope of human–animal relations, war is infrequently invoked as a relevant area of inquiry. War tends, after all, to call to mind intrahuman armed conflict among battling nation-states or powerful non-state entities. Although wars often do take such forms, they invariably involve animals in myriad ways. Animals have played key roles in major wars throughout history, including being used for the transport of soldiers and materials, to pass along communications, for medical assistance and locating wounded combatants, and to scout for weapons and carry out other forms of reconnaissance (see Cooper 1983 and Hediger 2013a for a historical overview of the role of animals in war). While the animals used in war are often exploited in ways that are ethically questionable, deep bonds have also been formed between soldiers and animals of many sorts, especially dogs (Frankel 2015; Lemish 1996).

In the United States, the military uses animals outside of armed combat for several kinds of research and experimentation, including psychological, medical, and weapons testing. The pro-animal organization People for the Ethical Treatment of Animals (PETA) has recently led a high-profile campaign against the military use of animals for medical purposes, particularly the use of so-called "live tissue" trauma training. In this sort of training, medics learn to treat traumatic injuries by working on live animals (such as goats and pigs) who have been deliberately injured to mimic human injuries encountered during war. PETA has shown that the animals used for these purposes are often inadequately anesthetized and suffer terribly from their wounds; further, PETA has made a compelling case that such training can be effectively carried out using human simulator models (PETA 2012).

Another important link between animals and war concerns the effects that armed conflicts have on animals' lives and habitats. Bombing, the use of mines, and the detonation of chemical weapons often injure and kill animals indiscriminately and sometimes destroy sensitive ecological regions. In cases where armed conflict leads to the sustained removal of human populations from a given habitat, wars do sometimes have positive consequences for animals (insofar as the absence of human populations sometimes allows nonhuman species to repopulate). More typically, though, conflicts (especially when they are frequent and extensive) cause serious declines in animal populations and further endanger already vulnerable individuals and species (Daskin and Pringle 2018).

Critical animal studies scholar Dinesh Wadiwel (2015) argues that we should go beyond examining the indirect effects of intrahuman war on animals to recognize that modern human societies are, in fact, *directly* at war with animals themselves. Although many people do not deliberately target animals with death or hold openly hostile attitudes toward most animals, Wadiwel argues that the basic foundations and structures of advanced industrialized societies are constituted in such a way as to be in a permanent state of war with animals. Although Wadiwel's position might seem counterintuitive, one need only consider the wide number of institutions and practices examined in this volume that involve the capture, control, torture, mutilation, injury, and **death** of countless numbers of animals to confirm his basic thesis (refer to the "*See also*" section at the end of this entry for a list of some of the relevant key concepts). If Wadiwel is correct about the widespread war against animals, then those who wish to address and contest such violence need to go beyond making individual ethical changes (which by themselves will not ultimately transform larger social structures) and become involved in explicitly political projects aimed at challenging institutions that claim sovereignty over animals and other marginalized populations (both human and **more-than-human**). Ultimately, Wadiwel argues that the first and most important step in trying to end the war against animals is to call a truce and to attempt, at least for brief periods, to halt the massive state and economic machinery used to abuse animals so that we can reflect on and begin to implement more pacific and respectful relations.

See also: biopolitics; CAFO; captivity; entertainment; experimentation; hunting; work

Further reading: Hediger 2013b; Kosek 2010

WILDLIFE

The dominant focus of animal **ethics** and animal studies has been on domesticated animals. Practitioners have sought to draw attention to prevailing attitudes and beliefs about the animals with which the vast majority of people frequently interact (see the entries on **speciesism** and **anthropocentrism**), while also challenging the unmitigated violence directed at domesticated animals (for example, in **CAFO**s or in medical testing and **experimentation**). But as various pro-animal frameworks have been further developed and refined over the past several decades, discussions have increasingly turned toward issues relating to wild animals and wildlife. Such concerns have arisen not simply from a concern with being conceptually and ethically consistent (as though being "pro-animal" logically entails a universal concern for the status of every entity labeled "animal"). Rather, the question of wildlife has emerged with special force in our time due to the heightened contact between human beings and wild animals in an ever-more-populated world and in relation to the variety of problems associated with ecological degradation and **climate change.** Today, when human and animal worlds interpenetrate in multiple ways, it is not enough for pro-animal discourse and activism to restrict itself to the reform and/or abolition of violent cultural institutions and assume that an ethic of non-interference will suffice in regard to our interactions with wildlife (Donaldson and Kymlicka 2011). Rather, changing planetary conditions are now making it necessary to address a whole range of human–wildlife relations, both those in which human beings affect wildlife (for example, through degrading the habitat of wild animals) and those in which wildlife intrudes upon the human world (for example, through the introduction of novel zoonotic diseases into human societies).

There are important precedents and extant policy frameworks for adjudicating some of the complex ethical and political dimensions of human impacts on wild animals. In the United States, the Endangered Species Act (ESA) is without doubt the most robust piece of legislation used to protect and respect the existence of wildlife. Established over the course of the 1960s and 1970s, the ESA takes as its charge the protection of species that are either on the brink of **extinction** or on the path toward becoming extinct. One of the remarkable features of the ESA is its mandate to protect not only endangered species themselves but the habitats on which such species rely for their continued existence (see Alagona 2013 for further analysis of the importance of habitat protection in the enforcement of the ESA). Despite its far-reaching vision, the ESA has led to unanticipated ethical paradoxes and faced considerable obstacles in its implementation. On the one hand, protection of endangered species has sometimes led to holding certain animals captive in order to ensure their continued existence and reproduction; wildlife managers have also deployed various apparatuses of control that seek to trap, track, surveil, and manage the movement of endangered animal species in ways that pose ethical concerns (Benson 2010). On the other hand, the ESA faces protracted and often effective opposition from powerful lobbyist organizations representing industries who wish to exploit endangered species and their habitats for commercial purposes. (See Baur and Irvin 2010 for a fuller discussion of the ESA and its various strengths and shortcomings.) Other popular strategies for protecting wildlife include the establishment of national parks and refuges, which are intended to provide a safe space for endangered species to repopulate and establish healthy ecosystems (see the entry on **rewilding** for more on this topic). Such wildlife spaces face opposition from the same industry forces that challenge the scope and implementation of the ESA as well as from marginalized populations who often rely on the land and animals in a given area for subsistence (Grebowicz 2015).

The ESA and related animal and environmental **laws** and policies are unquestionably important bulwarks for staving off the extinction of certain endangered species, but they do not constitute anything like a full framework for addressing human obligations toward wildlife or for the joint flourishing of human and wildlife communities. For even though species extinction and biodiversity loss are urgent problems, the vast majority of human–wildlife interactions occur in relation to species that do not fit within the purview of these frameworks. What is more, given the speed and scope of ecological changes in our era (see the entry on the **Anthropocene**), our most basic ideas about how to define and think about wildlife are undergoing concomitant transformations.

According to geographer Jamie Lorimer (2015), many of the current issues we face in regard to the human–wildlife relations stem from the use of outmoded conceptions of nature and wilderness. We have a tendency, Lorimer argues, to see nature and wilderness as purely nonhuman domains that either should be preserved as such (à la classical preservationist environmentalism) or occupied and exploited as much as possible for human gain (à la neoliberal versions of **capitalism**). Lorimer maintains, though, that such notions of pure nature and wilderness are now dead (if they were ever meaningful). In the Anthropocene era, the human footprint now reaches into every nook and cranny of the **more-than-human** world; conversely, despite expanding efforts at domination and subjection, the natural world remains beyond full human mastery. But if pure wilderness is no longer a tenable concept, perhaps *wildness* is. For Lorimer, the notion of wildness can help us do justice to wildlife and the more-than-human world in both a conceptual and ethico-political sense. Rather than encouraging us to focus primarily on saving individual animals and plants or distinct species, an emphasis on wildness turns our attention to the complex, relational, vital, and abundant forces that make human and more-than-human life as a whole possible. It is wildness that allows life to proliferate, take on ever new forms, and increase in biodiversity. Thus, if we wish to do justice to wildlife and improve human–wildlife relations, on Lorimer's account it would behove us to learn first and foremost to affirm and respect the creative and relational flows of wildness that we find both within ourselves and in the natural world.

See also: Anthropocene; becoming-animal; climate change; companion species; zoo; zoopolis

Further reading: Baur and Irvin 2010; Burgess 2001

WORK

Among human–animal relations, work is one of the most complicated; it is also one of the least studied. The latter fact is surprising given the ubiquity of working animals in past and present human societies. Animals perform and are involved with work for the purposes of agriculture, medical research, security, transportation, and **entertainment**, to name just a few of the more prominent industries and practices. (Coulter 2016 provides an extensive list of types of animal work and helpfully distinguishes between work done *with/for* animals and *by* animals.) What this list suggests is that there can be no single approach to the question of animal work, as its forms are too manifold to issue in a simple affirmation or critique.

Many animal rights theorists would point out, however, that the *vast* majority of the work done with and by animals is involuntary and exploitative—which is to say, in most cases animals have to be forcibly trained to do the work at hand and frequently receive little or no benefit from it. Consequently, animal rights advocates often push for the abolition of animal work in nearly all of its forms. Critical animal studies theorists such as Jason Hribal (2003) go further and call not simply for the abolition of animal work but for deeper reflection on the capitalist economic system that exploits the labor power of both animals and human beings. For Hribal, animal work should be seen as one kind of exploited labor among others. As with dispossessed workers, under a capitalist economic system animals are alienated from both their labor power and the commodities they produce. With these similarities in mind, Hribal argues that animals should be seen as members of the proletariat (the exploited working class within the capitalist economic system) and that animal liberation is fundamentally linked to the overthrow of capitalism and the liberation of all workers.

Other theorists such as Jocelyne Porcher have a less revisionist view of animal work. Although Porcher acknowledges that animals are often brutally exploited and killed in human–animal work relations, she argues that much of the work done by and with animals can be reformed in a more "peaceful and intelligent" direction (2017a: 304). Porcher believes that the abolitionist perspective leads to an undesirable future in which human beings will be largely removed from contact with animals in the name of avoiding exploitation and achieving personal ethical purity. To illustrate this point, take farming and animal agriculture as an example. On the abolitionist paradigm, not only would the practices of eating meat, eggs, and dairy be abolished, but the practice of using animals for their labor in the fields or around the farm would also be eliminated. In place of these animal-based operations, Porcher assumes liberationists would advocate for mechanized modes of agriculture and **meat** substitutes. If such alternatives were embraced, though, a number of worthwhile human–animal relations and bonds would be dissolved. Consequently, Porcher argues that we should seek to reform human–animal work on the basis of a fuller view of (1) how work is now central to many animals' lives, and (2) how deeply indebted human society is to the contributions of working animals.

Affirming central insights from both the abolitionist and reformist perspectives, Coulter (2016) argues that pro-animal activists and theorists should seek to develop more opportunities for what she calls *humane work*. For Coulter, humane jobs are ones "that are good for both people and animals" and that "prioritize both material and experiential wellbeing" (2016: 163). In view of the pressing need to create work that is socially, economically, and ecologically sustainable and meaningful, Coulter suggests that advocacy for humane jobs should constitute a central plank of contemporary animal studies.

See also: abolitionism; capitalism

Further reading: Coulter 2016; Hribal 2007; Noske 1989; Porcher 2017b; Hamilton and Taylor 2013

WORLD

The concept of world as it appears in animal studies typically derives from the writings of biologist Jakob von Uexküll (1864–1944). Responding to the dominance of reductive, mechanistic approaches to the study of animals in his time, Uexküll develops the concept of the animal *Umwelt*—that is, the animal's surrounding, lived, perceptual world—as the cornerstone of a more holistic framework for thinking about animal existence. Uexküll argues we can gain a fuller appreciation for the nature of a given animal's life if we learn to inhabit its world (von Uexküll 1985). He famously describes these worlds in metaphorical terms as "bubbles" that surround animals and make certain features of their environment accessible as they move through their surroundings (von Uexküll 2010). In familiarizing ourselves with a particular animal's world, we come to see the animal not simply as an *object* of scientific study but as a *subject* in its own right, who relates to its environment and other animals in meaningful ways. This kind of world-based, holistic approach to studying animals has two chief implications. On the one hand, it multiplies the number of meaningful "worlds" in an exponential manner and suggests that meaning can be found in multiple forms well beyond the boundaries of human culture. On the other hand, as philosopher Brett Buchanan argues, Uexküll's approach encourages the researcher to think of an individual animal's world as an *inter*-subjective sphere, since an animal's meaningful world is always constituted with and through others of various sorts (Buchanan 2008). This second aspect has profound consequences for how we think about both human–animal inter-subjectivity (that is, the shared worlds that human beings and animals co-constitute) and animal–animal intersubjectivity (that is, the shared worlds that animals build among themselves).

Uexküll's writings on world and related topics in animal biology have been developed in numerous directions and have had a sig-nificant influence on several academic fields (such as biology, biose-miotics, and **ethology**) and numerous leading intellectuals (including Ernst Cassirer, Maurice Merleau-Ponty, Gilles Deleuze, and Giorgio Agamben) (see Brentari 2015 for a survey of Uexküll's work and his intellectual influence and legacy). But it is philosopher Martin Heidegger who offers what is arguably the most influential response to the Uexküllian idea that animals inhabit meaningful worlds. In his magnum opus *Being and Time* (Heidegger 1962), Heidegger maintains that one of the defining features of human beings is that they exist and take up projects within a meaningful world in ways that animals and other nonhuman beings do not. In a lecture course delivered two years after *Being and Time*, Heidegger revisits the concept of world at more length in view of explicitly challenging Uexküll's idea that animals have world (Heidegger 1995). In these lectures Heidegger acknowledges that, at first glance, it does appear that animals have something like a world—if by "world" we mean having access to other beings in one's natural environment. But, Heidegger suggests, even if animals do have access to other beings in some way, that access appears impoverished when compared with the richly meaningful worlds human beings inhabit. As a way of underscoring this differ-ence, Heidegger posits his famous thesis that "animals are poor in world" (Heidegger 1995).

Ultimately, though, Heidegger's challenge to the notion that ani-mals inhabit worlds is much more thorough than this oft-mentioned thesis would suggest. Contesting Uexküll's position in the strongest possible terms, Heidegger maintains that in the final analysis animals do not, in fact, inhabit meaningful worlds *at all*. A meaningful world cannot, Heidegger insists, be equated simply with having some kind of mutual interaction with other beings in one's environment (which is what he believes is characteristic of animal life). Rather, a mean-ingful world becomes possible only for beings (like human beings) who have a deeply affective, relational, and free interaction with other entities. For Heidegger, the notion that only human beings have a world can be shown by (what he claims is) the fact that only human beings are able to notice *that* there are other beings, which is to say, only human beings notice beings *as* beings. Animals interact with other beings, to be sure, but they do so, Heidegger argues, while suspended entirely within the orbit of their biologically driven behavioral structures. Thus, he argues, animals are incapable of ever having those structures interrupted and hence can never notice *that* other beings exceed them. Heidegger's complicated and controversial analysis of animality and world been challenged by many animal philosophers (see Derrida 2008 for what is perhaps the most influential critical reading) and supported by other scholars (for example, Holland 2018).

Further reading: Buchanan 2008; Brentari 2015

ZOOPOLIS

Zoopolis (*zōon* "animal" + *polis* "city") is a concept meant to draw attention to the role that animals play in (and in relation to) predominantly human cities and urban environments. Traditionally, animal ethicists have directed the bulk of their attention to curbing the most visible forms of violence toward animals as they occur in institutions such as **CAFO**s and scientific labs. While these issues are undoubtedly important ones, focusing exclusively on such problems leaves unaddressed the status of animals outside of commercial and industrialized settings. In this regard, we might wonder about what we owe to urban animals, "undomesticated" animals, **wildlife**, marine animals, and so on. Should we aim for an ethic of non-interference with regard to such animals? Or would it be preferable to assume that encounters with such animals are inevitable and thus develop a set of practices and institutions that encourage respectful and joyful relations along these lines?

Jennifer Wolch (1998) uses the term zoopolis to refer to this latter option. She calls for pro-animal thinkers and activists to find ways to "renaturalize cities and invite the animals back in" (124). For Wolch, this process of renaturalization is done not just for the benefit of human beings (who, in re-encountering animals are reintroduced to the wonder and richness of animal life), but also for the benefit of animals themselves. Wolch argues that unless urban human beings relearn to interact with animals, they will never fully understand the sorts of personal and structural changes that need to be made in order to do justice to animal lives. For example, if we limit our focus simply to curbing meat consumption and buying products that are not tested on animals, we will fail to see the myriad ways in which dominant forms of urban life are themselves fundamentally at odds with the flourishing of animals. For Wolch, if we start from the perspective of a renaturalized way of life and a commitment to having caring encounters with animals, we might finally gain the awareness and inspiration to reform a whole range of practices that negatively impact animals, from driving to energy use to urban design. Wolch also suggests that proceeding from this stance allows pro-animal discourse and activism to rejoin movements for social justice that typically have their origins in urban environments (as opposed to say, a pro-animal approach that is grounded in a wilderness ethic divorced from urban life or that simply relocates urban and wild animals to sanctuaries or reserves outside of the city).

Although inspired by Wolch's vision for a zoopolis, Sue Donaldson and Will Kymlicka (2011) suggest that we need a less enchanted and more refined account of the status of animals in relation to cities and urban settings. Contra Wolch, Donaldson and Kymlicka argue that "we can take reasonable steps to keep would-be opportunistic animals out" (2011: 299, n. 46) of urban settings, and that a zoopolis does not require a full-scale renaturalization and re-enchantment of the city. Instead, they suggest that domestic animals, urban/liminal animals, and wild animals require different and specific kinds of political standing in relation to human communities. In view of setting up just relations with animals, they argue that domestic animals who live directly among human beings should be given citizenship; wild species and groups of animals who live outside of urban settings should be treated as independent and sovereign communities; and urban/liminal animals should be seen as denizens with partial citizenship rights insofar as their communities overlap with human communities. Thus, in regard to undomesticated animals who live near human cities, Donaldson and Kymlicka believe that we should aim to establish a political relationship "that is consistent with ... preserving relations of wariness and distrust" (2011: 299, n. 46). The political concept of denizenship, they maintain, better describes such relations of "co-residence" and avoids subsuming urban animals under the misleading rubric of "co-membership" in the same city or community.

Further reading: Michelfelder, 2003; Philo and Wilbert, 2000

ZOOS

Zoos are sites where wild animals are displayed for public viewing. Often included under this term are related establishments such as aquariums, oceanariums, wildlife parks, safari parks, and other public institutions that house wild animals. The zoos most visited today are relatively modern institutions; they were preceded by various practices of keeping wild animals that stretch back in history to the earliest human civilizations (Rothfels 2002). Historically, collections of wild animals were privately held by aristocrats; and menageries open to elite sectors of society were common throughout various historical periods and cultures. Some 200 years ago, zoological gardens (from which the modern word *zoo* is derived) emerged in Europe that made collections of wild animals available for study and eventually for public viewing. Zoos were not introduced in the United States until the last century, but they have grown rapidly in number and popularity since that time (Kisling 2001; Baratay and Hardouin-Fugier 2002). Today, there are an estimated 10,000 institutions across the world that qualify as zoos, with some 1,200 of those zoos being accredited by the World Association of Zoos and Aquariums. These 1,200 accredited zoos alone receive upwards of 700 million visitors a year, making zoos one of the most common sites of encounter between human beings and non-domesticated animals (Gusset and Dick 2011). The high numbers of people who visit zoos testify to the enduring and widespread public interest in and appreciation of **wildlife**; the question for researchers in animal studies is whether zoos are the best place for cultivating such encounters.

Modern zoos are primarily run as businesses and, as such, must make a profit by providing entertainment for visitors. This business model has led to the criticism that zoos are, at bottom, exploitative institutions that care more for profit than for the welfare or rights of the animals they house. In response to such criticism, many modern zoos have sought to create a new image of and rationale for zoos that emphasize their social and ecological benefits. Thus, many zoos now present themselves not simply as sites for public entertainment but as educational institutions that teach visitors about the importance and value of wild animals; there are, after all, very few places where contemporary urban human beings are able to have such close encounters with wild animals and to learn about them in a hands-on context. In addition, modern zoos argue that they provide refuge for animals on the verge of **extinction**, housing the few remaining individuals of a given species and allowing them to reproduce before eventually re-inhabiting the wild. And in cases where releasing a member of an endangered species back into its original habitat is no longer possible, many zoos claim they provide sanctuaries for such species to be conserved and cherished for as long as is reasonably possible. Following this alternative conception of what zoos can be, Jenny Gray, CEO of Zoos Victoria, makes the case that zoos can be ethical institutions if they are grounded in the principle of "compassionate conservation" (Gray 2017). For Gray, such compassion entails making animal welfare a central concern of the day-to-day operations of zoos and placing conservation at the heart of their overall mission. Yet, that same ethical principle, she notes, also means rejecting certain traditional zoo practices (such as capturing wild animals who are not endangered) in favor of more ethical means of keeping animals.

Many **animal welfare and animal rights** advocates are unpersuaded by such arguments in defense of zoos. They argue, for instance, that zoos make exaggerated claims about the positive effects of their education outreach as well as the importance of their role in species preservation (Jamieson 1985). Critics further argue that, even if zoos were designed and maintained such that animal welfare were a priority, simply keeping animals in **captivity** raises a host of ethical problems. At a minimum, from an animal rights perspective keeping animals in zoos would seem to violate their basic right to pursue their interests autonomously (Regan 1995). That many zoo animals attempt to escape their enclosures, suffer from physical and psychological stress, and often have contentious relationships with their keepers and visitors all indicate that zoos are imposing a mode of existence on animals that disrespects their most fundamental interests and **agency** (Hribal 2010).

Beyond concerns with the rights of animals, animal studies theorists have also emphasized the profoundly negative effects that zoos can have on the mindsets and subjectivities of human visitors (Acampora 1998; Chrulew 2011). In particular, if we appreciate that zoos as architectural apparatuses that configure relations among human beings and animals such that animals are encountered solely as objects for human spectators, it becomes clear that zoos are hardly neutral sites for encounter. In fact, as critics maintain, they are one of the chief sites where **anthropocentrism** and human supremacy are reinforced and where human beings are actually *denied* meaningful encounters with animals in all of their otherness and difference. For animal studies theorists who adopt this perspective, the point is not simply to abolish zoos but to generate other kinds of encounters with wildlife that are more respectful and meaningful and that generate new possibilities for living well with our fellow animals.

Further reading: Braverman 2013; Hosey, Melfi, and Pankhurst 2013; Minteer, Maienschein, and Collins 2018; Safina 2018

BIBLIOGRAPHY

Abram, D. (1996) *The Spell of the Sensuous: Perception and Language in a More-Than-Human World*, New York: Pantheon.

Abram, D. (2005) "Between the Body and the Breathing Earth: A Reply to Ted Toadvine," *Environmental Ethics*, 27: 171–190.

Abram, D. (2010) *Becoming Animal: An Earthly Cosmology*, New York: Pantheon.

Acampora, R. (1998) "Extinction by Exhibition: Looking at and in the Zoo," *Human Ecology Review*, 5: 1–4.

Adams, C. J. (2015 [1990]) *The Sexual Politics of Meat: A Feminist-Vegetarian Critical Theory*, New York: Bloomsbury.

Adams, C. J. (2004) *The Pornography of Meat*, New York: Continuum.

Adams, C. J. (ed.) (2016) *The Carol J. Adams Reader: Writings and Conversations, 1995–2015*, New York: Bloomsbury.

Adams, C. J. and Donovan, J. (eds.) (1995) *Animals and Women: Feminist Theoretical Explorations*, Durham, NC: Duke University Press.

Adams, C. J. and Gruen, L. (eds.) (2014) *Ecofeminism: Feminist Intersections with Other Animals and the Earth*, New York: Bloomsbury.

Agamben, G. (1998) *Homo Sacer: Sovereign Power and Bare Life*, tr. D. Heller-Roazen, Stanford, CA: Stanford University Press.

Agamben, G. (2004) *The Open: Man and Animal*, tr. Kevin Attell, Stanford, CA: Stanford University Press.

Agamben, G. (2014) *The Use of Bodies*, tr. Adam Kotsko, Stanford, CA: Stanford University Press.

Alagona, P. S. (2013). *After the Grizzly: Endangered Species and the Politics of Place in California*, Berkeley: University of California Press.

Alaimo, S. (2010) "Eluding Capture: The Science, Culture, and Pleasure of 'Queer' Animals," in C. Mortimer-Sandilands and B. Erickson (eds.), *Queer Ecologies: Sex, Nature, Politics, Desire*, Bloomington, IN: Indiana University Press, 51–72.

Alaimo, S. (2016) *Exposed: Environmental Politics and Pleasures in Posthuman Times*, Minneapolis: University of Minnesota Press.

Alaimo, S. and Hekman, S. (eds.) (2008) *Material Feminisms*, Bloomington, IN: Indiana University Press.

Aloi, G. (2012) *Art and Animals*, New York: I. B. Tauris.

Alvaro, C. (2019) "Lab-Grown Meat and Veganism: A Virtue-Oriented Perspective," *Journal of Agricultural and Environmental Ethics*, *32*: 127–141.

Anderson, V. D. (2004) *Creatures of Empire: How Domestic Animals Transformed Early America*, New York: Oxford University Press.

Andrzejewski, J., Pedersen, H., and Wicklund, F. (2009) "Interspecies Education for Humans, Animals, and the Earth," in J. Andrzejewski, M. Baltodano, and L. Symcos (eds.), *Social Justice, Peace, and Environmental Education: Transformative Standards*, New York: Routledge, 136–154.

APPA (American Pet Products Association) (2019) *"Pet Industry Market Size & Ownership Statistics,"* [Online] Available at: [https://www.americanpetproducts.org/press_industrytrends.asp], accessed 15 August 2019.

Archer, D. (2012) *Global Warming: Understanding the Forecast*, Hoboken, NJ: Wiley.

Arluke, A. (2003) "The No-Kill Controversy: Manifest and Latent Sources of Tension," in D. J. Salem and A. N. Rowan (eds.), *The State of the Animals II: 2003*, Washington, DC: Humane Society Press, 67–83.

Armstrong, S. J. and Botzler, R. G. (eds.) (2008) *The Animal Ethics Reader*, New York: Routledge.

Arnaquq-Baril, A. (2016) *Angry Inuk*, New York: Film Movement, DVD.

Asdal, K., Druglitrö, T., and Hinchliffe, S. (eds.) (2017) *Humans, Animals, and Biopolitics: The More-than-Human Condition*, London: Routledge.

AVMA (American Veterinary Medical Association) (2018) *U.S. Pet Ownership and Demographics Sourcebook*, Schaumburg, IL: AVMA.

Bagemihl, B. (1999) *Biological Exuberance: Animal Homosexuality and Natural Diversity*, New York: St. Martin's Press.

Bain, P. G., Vaes, J., and Leyens, J.-P. (eds.) (2014) *Humanness and Dehumanization*, New York: Psychology Press.

Baker, S. (2001) *Picturing the Beast: Animals, Identity, and Representation*, Champaign, IL: University of Illinois Press.

Baker, S. (2013) *Artist/Animal*, Minneapolis: University of Minnesota Press.

Balcombe, J. (2010) *Second Nature: The Inner Lives of Animals*, New York: Palgrave Macmillan.

Barad, K. (2007) *Meeting the Universe Halfway*, Durham, NC: Duke University Press.

Baratay, É. and Hardouin-Fugier, E. (2002) *Zoo: A History of Zoological Gardens in the West*, London: Reaktion.

Barnosky, A. D., Matzke, N., and Tomiya, S. et al. (2011) "Has the Earth's Sixth Mass Extinction Already Arrived?" *Nature*, *471*: 51–57.

Bastian, B., Costello, K., Loughnan, S. et al. (2012) "When Closing the Human–Animal Divide Expands Moral Concern: The Importance of Framing," *Social Psychological and Personality Science*, *3*: 421–429.

Bastian, M., Jones, O., Moore, N. et al. (eds.) (2016) *Participatory Research in More-than-Human Worlds*, London: Routledge.

Bataille, G. (2005) *The Cradle of Humanity: Prehistoric Art and Culture*, tr. Michelle Kendall and Stuart Kendall, New York: Zone Books.

Bateson, P. and Laland, K. N. (2013) "Tinbergen's Four Questions: An Appreciation and an Update," *Trends in Ecology & Evolution*, *28*: 712–718.

Bateson, P. and Martin, P. (2013) *Play, Playfulness, Creativity, and Innovation*, New York: Cambridge University Press.

Baur, D. C. and Irvin, W. R. (eds.) (2010) *Endangered Species Act: Law, Policy, and Perspectives*, Chicago, IL: American Bar Association.

Baur, G. (2008) *Farm Sanctuary: Changing Hearts and Minds about Animals and Food*, New York: Simon & Schuster.

Beaulieu, A. (2011) "The Status of Animality in Deleuze's Thought," *Journal for Critical Animal Studies*, *9*: 69–88.

Beck, A. and Katcher, A. (1996) *Between Pets and People: The Importance of Animal Companionship*, West Lafayette, IN: Purdue University Press.

Beetz, A.M. (2008) "Bestiality and Zoophilia: A Discussion of Sexual Contact with Animals," in F. R. Ascione (ed.), *The International Handbook of Animal Abuse and Cruelty: Theory, Research and Application*, West Lafayette, IN: Purdue University Press, 201–220.

Beirne, P. (2009) *Confronting Animal Abuse: Law, Criminology, and Human–Animal Relationships*, Lanham, MD: Rowman & Littlefield.

Bekoff, M. (1998) "Deep Ethology, Animal Rights, and the Great Ape/Animal Project: Resisting Speciesism and Expanding the Community of Equals," *Journal of Agricultural and Environmental Ethics*, *10*: 269–296.

Bekoff, M. (2007) *The Emotional Lives of Animals: A Leading Scientist Explores Animal Joy, Sorrow, and Empathy and Why They Matter*, Novato, CA: New World Library.

Bekoff, M. (2014) *Rewilding Our Hearts: Building Pathways of Compassion and Coexistence*, Novato, CA: New World Library.

Bekoff, M. and Byers, J. A. (eds.) (1998) *Animal Play: Evolutionary, Comparative, and Ecological Perspectives*, New York: Cambridge University Press.

Bekoff, M. and Pierce, J. (2009) *Wild Justice: The Moral Lives of Animals*, Chicago, IL: University of Chicago Press.

Bekoff, M., Allen, C., and Burghardt, G. M. (eds.) (2002) *The Cognitive Animal: Empirical and Theoretical Perspectives on Animal Cognition*, Cambridge, MA: MIT Press.

Belcourt, B.-R. (2014) "Animal Bodies, Colonial Subjects: (Re)Locating Animality in Decolonial Thought," *Societies*, *5*: 1–11.

Benatar, D. (2008). *Better Never to Have Been: The Harm of Coming into Existence*, New York: Oxford University Press.

Bennett, J. (2010) *Vibrant Matter: A Political Ecology of Things*, Durham, NC: Duke University Press.

Benson, E. (2010) *Wired Wilderness: Technologies of Tracking and the Making of Modern Wildlife*, Baltimore, MD: Johns Hopkins University Press.

Benton, T. (1993) *Natural Relations: Ecology, Animal Rights and Social Justice*, London: Verso.

Best, S. (2014) *The Politics of Total Liberation: Revolution for the 21st Century*, New York: Palgrave Macmillan.

Best, S. and Nocella, A. J. (eds.) (2004) *Terrorists or Freedom Fighters?: Reflections on the Liberation of Animals*, New York: Lantern Books.

Best, S. and Nocella, A. J. (eds.) (2006) *Igniting a Revolution: Voices in Defense of the Earth*, Oakland, CA: AK Press.

Birke, L. and Holmber, T. (2018) "Intersections: The Animal Question Meets Feminist Theory," in C. Åsberg and R. Braidotti (eds.), *A Feminist Companion to the Posthumanities*, Cham: Springer, 117–128.

Birke, L., Arluke, A. and Michael, M. (eds.) (2007) *The Sacrifice: How Scientific Experiments Transform Animals and People*, West Lafayette, IN: Purdue University Press.

Blunden, J., Arndt, D. S., and Hartfield, G. (eds.) (2018) *"State of the Climate in 2017,"* Bulletin of the American Meteorological Society, 99: Si–S332.

Boisseron, B. (2018) *Afro-Dog: Blackness and the Animal Question*, New York: Columbia University Press.

Bradbury, J. W. and Vehrencamp, S. L. (2011) *Principles of Animal Communication*, Sunderland, MA: Sinauer.

Braidotti, R. (2013) *The Posthuman*, Malden, MA: Polity Press.

Braithwaite, V. (2010) *Do Fish Feel Pain?*, New York: Oxford University Press.

Braverman, I. (2013) *Zooland: The Institution of Captivity*, Stanford: Stanford University Press.

Brentari, C. (2015) *Jakob von Uexküll: The Discovery of the Umwelt between Biosemiotics and Theoretical Biology*, Dordrecht: Springer.

Broglio, R. (2011) *Surface Encounters: Thinking with Animals and Art*, Minneapolis: University of Minnesota Press.

Broom, D. M. (2014) *Sentience and Animal Welfare*, Wallingford, UK: CABI.

Buchanan, B. (2008) *Onto-Ethologies: The Animal Environments of Uexküll, Heidegger, Merleau-Ponty, and Deleuze*, Albany: SUNY Press.

Budiansky, S. (1992) *The Covenant of the Wild: Why Animals Chose Domestication*, New York: W. Morrow.

Burgess, B. B. (2001) *Fate of the Wild: The Endangered Species Act and the Future of Biodiversity*, Athens, GA: University of Georgia Press.

Burghardt, G. M. (1997) "Amending Tinbergen: A Fifth Aim for Ethology," in R. W. Mitchell, N. Thompson, and L. Miles (eds.), *Anthropomorphism, Anecdotes, and Animals*, Albany, NY: SUNY Press, 254–276.

Burghardt, G. M. (2005) *The Genesis of Animal Play: Testing the Limits*, Cambridge, MA: MIT Press.

Burghardt, G. M. (2015) "Play in Fishes, Frogs, and Reptiles," *Current Biology*, 25: R9–R10.

Burkhardt, R. W. (2005) *Patterns of Behavior: Konrad Lorenz, Niko Tinbergen, and the Founding of Ethology*, Chicago, IL: University of Chicago Press.

Butler, J. (2004) *Precarious Life: The Powers of Mourning and Violence*, New York: Verso.

Butler, J. (2009) *Frames of War: When is Life Grievable?*, London: Verso.

Calarco, M. (2008) *Zoographies: The Question of the Animal from Heidegger to Derrida*, New York: Columbia University Press.

Calarco, M. (2015) *Thinking Through Animals: Identity, Difference, Indistinction*, Stanford, CA: Stanford University Press.

Callicott, J. B. (1980) "Animal Liberation: A Triangular Affair," *Environmental Ethics*, 2: 311–328.

Carlson, M. (2018) *Affect, Animals, and Autists: Feeling around the Edges of the Human in Performance*, Ann Arbor, MI: University of Michigan Press.

Carruthers, P. (1992) *The Animals Issue: Moral Theory in Practice*, New York: Cambridge University Press.

Carter, B. and Charles, N. (2013) "Animals, Agency, and Resistance," *Journal for the Theory of Social Behavior, 43*: 322–340.

Cassidy, R. (2009) "Zoosex and other Relationships with Animals," in H. Donnan and F. Magowan (eds.), *Transgressive Sex: Subversion and Control in Erotic Encounters*, New York: Berghahn, 91–112.

Causey, A. S. (1989) "On the Morality of Hunting," *Environmental Ethics, 11*: 327–343.

Cavalieri, P. (2001) *The Animal Question: Why Nonhuman Animals Deserve Human Rights*, tr. Catherine Woollard, New York: Oxford University Press.

Cavalieri, P. (ed.) (2016) *Philosophy and the Politics of Animal Liberation*, New York: Palgrave Macmillan.

Ceballos, G., Ehrlich, P., Barnosky, A. et al. (2015) "Accelerated Modern Human-Induced Species Losses: Entering the Sixth Mass Extinction," *Science Advances, 1*: 1–5.

Chaudhuri, U. and Hughes, H. (eds.) (2014) *Animal Acts: Performing Species Today*, Ann Arbor, MI: University of Michigan Press.

Chen, M. Y. (2012) *Animacies: Biopolitics, Racial Mattering, and Queer Affect*, Durham, NC: Duke University Press.

Chew, M. and Carroll, S. (2011) *"The Invasive Ideology: Biologists and Conservationists Are Too Eager to Demonize Non-Native Species,"* *The Scientist* [Online] Available at: [https://www.the-scientist.com/news-opinion/opinion-the-invasive-ideology-41967], accessed 12 September 2018.

Chrulew, M. (2011) *"Managing Love and Death at the Zoo: The Biopolitics of Endangered Species Preservation,"* *Australian Humanities Review* [Online] Available at: [http://australianhumanitiesreview.org/2011/05/01/managing-love-and-death-at-the-zoo-the-biopolitics-of-endangered-species-preservation/], accessed 12 September 2018.

Chrulew, M. (2012) "Animals in Biopolitical Theory: Between Agamben and Negri," *New Formations, 76*: 53–67.

Chrulew, M. and De Vos, R. (2018) "Extinction," in L. Turner, R. Broglio, and U. Sellbach (eds.), *The Edinburgh Companion to Animal Studies*, Edinburgh: Edinburgh University Press, 181–197.

Cianchi, J. (2015) *Radical Environmentalism: Nature, Identity and More-than-Human Agency*, New York: Palgrave Macmillan.

Clark, E. (2012) "'The Animal' and 'The Feminist,'" *Hypatia*, 27: 516–520.

Clark, J. (2014) "Laborers or Lab Tools? Rethinking the Role of Lab Animals in Clinical Trials," in N. Taylor and R. Twine (eds.), *The Rise of Critical Animal Studies: From the Margins to the Center*, London: Routledge, 139–164.

Clark, J. (2017) *"Consider the Vulture: An Ethical Approach to Roadkill," Discard Studies* [Online] Available at: [https://discardstudies.com/2017/12/04/consider-the-vulture-an-ethical-approach-to-roadkill/], accessed 12 September 2018.

Clark, S. R. L. (1999) *The Political Animal: Biology, Ethics and Politics*, London: Routledge.

Clarke, A. E. and Haraway, D. J. (eds.) (2018) *Making Kin Not Population: Reconceiving Generations*, Chicago, IL: Prickly Paradigm Press.

Coates, P. (2007) *American Perceptions of Immigrant and Invasive Species: Strangers on the Land*, Berkeley: University of California Press.

Cochrane, A. (2014) "Born in Chains? The Ethics of Animal Domestication," in L. Gruen (ed.), *The Ethics of Captivity*, New York: Oxford University Press, 156–173.

Coe, S. (1995) *Dead Meat*, New York: Four Walls Eight Windows.

Cooper, J. (1983) *Animals in War*, London: Corgi.

Cordeiro-Rodrigues, L. and Mitchell, L. (eds.) (2017) *Animals, Race, and Multiculturalism*, New York: Palgrave Macmillan.

Corlett, R. T. (2016) "Restoration, Reintroduction, and Rewilding in a Changing World," *Trends in Ecology & Evolution*, 31: 453–462.

Costello, K. and Hodson, G. (2009) "Exploring the Roots of Dehumanization: The Role of Animal–Human Similarity in Promoting Immigrant Humanization," *Group Processes & Intergroup Relations*, 13: 3–22.

Coulter, K. (2016) *Animals, Work, and the Promise of Interspecies Solidarity*, New York: Palgrave Macmillan.

Council Directive 86/609/EEC (1986) *"On the Approximation of Laws, Regulations and Administrative Provisions of the Member States Regarding the Protection of Animals Used for Experimental and Other Scientific Purposes,"* [Online] Available at: [https://eur-lex.europa.eu/legal-content/EN/ALL/?uri=CELEX%3A31986L0609], accessed 12 September 2018.

Crary, A. (2019) "The Horrific History of Comparisons between Cognitive Disability and Animality (And How to Move Past It)," in L. Gruen and F. Probyn-Rapsey (eds.), *Animaladies: Gender, Animals, and Madness*, New York: Bloomsbury, 117–136.

Crenshaw, K. (1989) "Demarginalizing the Intersection of Race and Sex: A Black Feminist Critique of Antidiscrimination Doctrine, Feminist Theory, and Antiracist Politics," *The University of Chicago Legal Forum, 140*: 139–167.

Crist, E. (1999) *Images of Animals: Anthropomorphism and Animal Mind*, Philadelphia, PA: Temple University Press.

Crist, E. (2013a) "Ecocide and the Extinction of Animal Minds," in M. Bekoff, *Ignoring Nature No More: The Case for Compassionate Conservation*, Chicago, IL: University of Chicago Press, 45–61.

Crist, E. (2013b) "On the Poverty of Our Nomenclature," *Environmental Humanities*, 3: 129–147.

Crist, E. (2018) "Anthropocentrism," in N. Castree, M. Hulme, and J. D. Proctor (eds.), *Companion to Environmental Studies*, New York: Routledge, 735–739.

Crist, E. (2019) *Abundant Earth: Toward an Ecological Civilization*, Chicago, IL: University of Chicago Press.

Cronin, J. K. (2018) *Art for Animals: Visual Culture and Animal Advocacy, 1870–1914*, University Park, PA: Penn State University Press.

Crutzen, P. J. and Stoermer, E. (2000) "The Anthropocene," *Global Change Newsletter (International Geosphere-Biosphere Programme)*, 41: 12–13.

Cudworth, E. (2011) "Climate Change, Industrial Animal Agriculture, and Complex Inequalities," *The International Journal of Science in Society*, 2: 323–334.

Cudworth, E. (2014) "Beyond Speciesism: Intersectionality, Critical Sociology and the Human Domination of Other Animals," in N. Taylor and R. Twine (eds.), *The Rise of Critical Animal Studies: From the Margins to the Center*, London: Routledge, 19–35.

Cull, L. (2011) *Theatres of Immanence: Deleuze and the Ethics of Performance*, New York: Palgrave Macmillan.

Darwin, C. (1964 [1859]) *On the Origin of Species*, Cambridge, MA: Harvard University Press.

Darwin, C. (1981 [1871]) *The Descent of Man, and Selection in Relation to Sex*, Princeton, NJ: Princeton University Press.

Darwin, C. (1998 [1872]) *The Expression of the Emotions in Man and Animals*, New York: Oxford University Press.

Daskin, J. H. and Pringle, R. M. (2018) "Warfare and Wildlife Declines in Africa's Protected Areas," *Nature*, 553: 328–332.

Daston, L. and Mitman, G. (eds.) (2005) *Thinking with Animals: New Perspectives on Anthropomorphism*, New York: Columbia University Press.

Davis, M. A., Chew, M. K., Hobbs, R. J. et al. (2011) "Don't Judge Species on their Origins," *Nature*, 474: 153–154.

Dawkins, M. (1980) *Animal Suffering: The Science of Animal Welfare*, London: Chapman and Hall.

Dawkins, R. (1993) "Gaps in the Mind," in P. Cavalieri and P. Singer (eds.), *The Great Ape Project: Equality beyond Humanity*, New York: St. Martin's Press, 80–87.

De Vos, J. M., Joppa, L. N., Gittleman, J. L. et al. (2015), "Estimating the Normal Background Rate of Species Extinction," *Conservation Biology*, 29: 452–462.

de Waal, F. (1999) "Anthropomorphism and Anthropodenial: Consistency in Our Thinking about Humans and Other Animals," *Philosophical Topics*, 27: 255–280.

de Waal, F. (2006) *Primates and Philosophers: How Morality Evolved*, Princeton, NJ: Princeton University Press.

de Waal, F. (2009) *The Age of Empathy: Nature's Lessons for a Kinder Society*, New York: Harmony Books.

de Waal, F. (2013) *The Bonobo and the Atheist: In Search of Humanism among the Primates*, New York: W. W. Norton.

de Waal, F. (2016) *Are We Smart Enough to Know How Smart Animals Are?*, New York: W. W. Norton.

de Waal, F. (2019) *Mama's Last Hug: Animal Emotions and What They Tell Us about Ourselves*, New York: W. W. Norton.

Deckha, M. (2012) "Toward a Postcolonial, Posthumanist Feminist Theory: Centralizing Race and Culture in Feminist Work on Nonhuman Animals," *Hypatia*, 27: 527–545.

Deckha, M. (2018) "Postcolonial," in L. Gruen (ed.), *Critical Terms for Animal Studies*, Chicago, IL: University of Chicago Press, 280–293.

Deckha, M. and Pritchard, E. (2016) "Recasting Our 'Wild' Neighbours: Toward an Egalitarian Legal Approach to Urban Wildlife Conflicts," *UBC Law Review*, 49: 161–202.

DeGrazia, D. (1996) *Taking Animals Seriously*, New York: Cambridge University Press.

Dekkers, M. (1994) *Dearest Pet: On Bestiality*, tr. P. Vincent, London: Verso.

Deleuze, G. (2003) *Francis Bacon: The Logic of Sensation*, tr. D. Smith, London: Continuum.

Deleuze, G. and Guattari, F. (1987) *A Thousand Plateaus: Capitalism and Schizophrenia*, tr. B. Massumi, Minneapolis: University of Minnesota Press.

Deleuze, G. and Guattari, F. (1994) *What is Philosophy?*, tr. H. Tomlinson and G. Burchell, New York: Columbia University Press.

Dell'Aversano, C. (2010) "The Love Whose Name Cannot Be Spoken: Queering the Human–Animal Bond," *Journal for Critical Animal Studies*, 8: 73–125.

DeMello, M. (2012) *Animals and Society: An Introduction to Human–Animal Studies*, New York: Columbia University Press.

DeMello, M. (2017) "Shelters and Sanctuaries," in C. L. Johnston and J. Urbanik (eds.), *Humans and Animals: A Geography of Coexistence*, Santa Barbara, CA: ABC-CLIO, 299–301.

DeMello, M. (ed.) (2010) *Teaching the Animal: Human–Animal Studies across the Disciplines*, New York: Lantern.

DeMello, M. (ed.) (2016) *Mourning Animals: Rituals and Practices Surrounding Animal Death*, East Lansing, MI: Michigan State University Press.

Derrida, J. (1995) *Aporias: Dying—Awaiting (One Another at) the "Limits of Truth"*, tr. Thomas Dutoit, Stanford, CA: Stanford University Press.

Derrida, J. (2008) *The Animal That Therefore I Am*, tr. David Wills, New York: Fordham University Press.

Derrida, J. (2009) *The Beast and the Sovereign*, tr. Geoffrey Bennington, Chicago, IL: University of Chicago Press.

Derrida, J. and Nancy, J.-L. (1991) "'Eating Well,' or the Calculation of the Subject: An Interview with Jacques Derrida," in E. Cadava, P. Connor, and J.-L. Nancy (eds.), *Who Comes After the Subject?* New York: Routledge, 96–119.

Derrida, J. and Roudinesco, E. (2004) *For What Tomorrow: A Dialogue*, tr. Jeff Fort, Stanford, CA: Stanford University Press.

Descola, P. (2013) *Beyond Nature and Culture*, tr. J. Lloyd, Chicago, IL: University of Chicago Press.

Desmond, J. (2016) *Displaying Death and Animating Life: Human–Animal Relations in Art, Science, and Everyday Life*, Chicago, IL: University of Chicago Press.

Despret, V. (2016) *What Would Animals Say if We Asked the Right Questions?* tr. B. Buchanan, Minneapolis: University of Minnesota Press.

Despret, V. (2017) "It Is an Entire World That Has Disappeared," in D. B. Rose, T. van Dooren, and M. Chrulew (eds.), *Extinction Studies: Stories of Time, Death and Generations*, New York: Columbia University Press, 217–222.

Dinker, K. and Pedersen, H. (2016) "Critical Animal Pedagogies: Re-learning our Relations with Animal Others," in H. Lees and N. Noddings (eds.), *Palgrave Handbook of Alternative Education*, London: Palgrave Macmillan, 415–430.

Donaldson, B. (2015) *Creaturely Cosmologies: Why Metaphysics Matters for Animal and Planetary Liberation*, Lanham, MD: Lexington.

Donaldson, B. and Carter, C. (eds.) (2016) *The Future of Meat Without Animals*, Lanham, MD: Rowman & Littlefield.

Donaldson, B. and King, A. (eds.) (2019) *Feeling Animal Death: Being Host to Ghosts*, Lanham, MD: Rowman & Littlefield.

Donaldson, S. and Kymlicka, W. (2011) *Zoopolis: A Political Theory of Animal Rights*, New York: Oxford University Press.

Donlan, C. J., Berger, J., Bock, C. E. et al. (2006) "Pleistocene Rewilding: An Optimistic Agenda for Twenty-First Century Conservation," *The American Naturalist*, 168: 660–681.

Donner, W. (1997) "Animal Rights and Native Hunters: A Critical Analysis of Wenzel's Defense," in A. Wellington, A. Greenbaum, and W. Cragg (eds.), *Canadian Issues in Environmental Ethics*, Peterborough, ON: Broadview Press, 153–184.

Donovan, J. (1990) "Animal Rights and Feminist Theory," *Signs*, 15: 350–375.

Donovan, J. (2006) "Feminism and the Treatment of Animals: From Care to Dialogue," *Signs*, 31: 305–329.

Donovan, J. and Adams, C. (eds.) (1996) *Beyond Animal Rights: A Feminist Caring Ethic for the Treatment of Animals*, New York: Continuum.

Dunayer, J. (2013) "The Rights of Sentient Beings: Moving Beyond Old and New Speciesism," in R. Corbey and A. Lanjouw (eds.) *The Politics of Species: Reshaping Our Relationships with Other Animals*, Cambridge: Cambridge University Press, 27–39.

Edelman, L. (2004) *No Future: Queer Theory and the Death Drive*, Durham, NC: Duke University Press.

Emel, J. and Neo, H. (eds.) (2015). *Political Ecologies of Meat*, New York: Routledge.

Emmerman, K. S. (2014) "Sanctuary, Not Remedy: The Problem of Captivity and the Need for Moral Repair," in L. Gruen (ed.), *The Ethics of Captivity*, New York: Oxford University Press, 231–247.

EPA (Environmental Protection Agency) (2007) *"Regulatory Definitions of Large CAFOs, Medium CAFO, and Small CAFOs,"* [Online] Available at: [https://www3.epa.gov/npdes/pubs/sector_table.pdf], accessed 12 September 2018.

Evans, V. (2014) *The Language Myth: Why Language Is Not an Instinct*, New York: Cambridge University Press.

Fagen, R. (1981) *Animal Play Behavior*, New York: Oxford University Press.

Fanon, F. (1967a) *Black Skin, White Masks*, tr. C. L. Markmann, New York: Grove.

Fanon, F. (1967b) *Towards the African Revolution*, tr. H. Chevalier, New York: Grove.

Favre, D. S. (2011) *Animal Law: Welfare, Interests, and Rights*, New York: Wolters Kluwer.

Few, M. and Tortorici, Z. (eds.) (2013) *Centering Animals in Latin American History*, Durham, NC: Duke University Press.

Fiddes, N. (1991) *Meat: A Natural Symbol*, New York: Routledge.

Fine, A. H. (ed.) (2019) *Handbook on Animal-Assisted Therapy: Foundations and Guidelines for Animal-Assisted Interventions*, San Diego, CA: Elsevier.

Finsen, L. and Finsen, S. (1994) *The Animal Rights Movement in America: From Compassion to Respect*, New York: Twayne.

Foote, J. (1990) "Trying to Take Back the Planet," *Newsweek, 115*: 24.

Foreman, D. (2004) *Rewilding North America: A Vision for Conservation in the 21st Century*, Washington, DC: Island Press.

Foucault, M. (1990) *The History of Sexuality*, Volume 1, tr. R. Hurley, New York: Vintage.

Foucault, M. (2003) *'Society Must Be Defended': Lectures at the Collège de France, 1975–1976*, tr. D. Macey, New York: Picador.

Francione, G. (1995) *Animals, Property, and the Law*, Philadelphia, PA: Temple University Press.

Francione, G. (1996) *Rain without Thunder: The Ideology of the Animal Rights Movement*, Philadelphia, PA: Temple University Press.

Francione, G. (2008) *Animals as Persons: Essays on the Abolition of Animal Exploitation*, New York: Columbia University Press.

Francione, G. L. and Charlton, A. E. (2016) *"The Case Against Pets,"* AEON *Magazine* [Online] Available at: [https://aeon.co/essays/why-keeping-a-pet-is-fundamentally-unethical], accessed 15 August 2019.

Francione, G. L. and Garner, R. (2010) *The Animal Rights Debate: Abolition or Regulation?*, New York: Columbia University Press.

Frankel, R. (2015) *War Dogs: Tales of Canine Heroism, History, and Love*, New York: St. Martin's.

Freud, S. (1955) "A Difficulty in the Path of Psycho-analysis," in J. Strachey (ed.), *The Standard Edition of the Complete Psychological Works of Sigmund Freud*, Vol. 17, London: Hogarth, 137–144.

Freund, P. and Martin, G. (2007) "Hyperautomobility, the Social Organization of Space, and Health," *Mobilities*, 2: 37–49.

Gaard, G. (2001) "Tools for a Cross Cultural Feminist Ethics: Exploring Ethical Contexts and Contents in the Makah Whale Hunt," *Hypatia*, 16: 1–26.

Gaard, G. (2012) "Speaking of Animal Bodies," *Hypatia*, 27: 521–526.

Gaard, G. (2017) "Posthumanism, Ecofeminism, and Inter-Species Relations," in S. MacGregor (ed.), *Routledge Handbook of Gender and Environment*, London: Routledge, 464–472.

Gaarder, E. (2011) *Women and the Animal Rights Movement*, New Brunswick, NJ: Rutgers University Press.

Gabardi, W. (2017) *The Next Social Contract: Animals, the Anthropocene, and Biopolitics*, Philadelphia, PA: Temple University Press.

Gane, N. and Haraway, D. J. (2006) "When We Have Never Been Human, What Is to Be Done?," *Theory, Culture & Society*, 23: 135–158.

Gardner, C. and MacCormack, P. (eds.) (2017) *Deleuze and the Animal*, Edinburgh: Edinburgh University Press.

Garner, R. (2005) *Animal Ethics*, Malden, MA: Polity.

Garrett, J. (ed.) (2012) *The Ethics of Animal Research: Exploring the Controversy*, Cambridge, MA: MIT Press.

Gerber, P.J., Steinfeld, H., Henderson, B. et al. (2013) *Tackling Climate Change through Livestock: A Global Assessment of Emissions and Mitigation Opportunities.* Rome: Food and Agriculture Organization of the United Nations.

GFK (Growth from Knowledge) (2016) *"Man's Best Friend: Global Pet Ownership and Feeding Trends,"* [Online] Available at: [https://www.gfk.com/insights/news/mans-best-friend-global-pet-ownership-and-feeding-trends/], accessed 15 August 2019.

Giffney, N. and Hird, M. J. (eds.) (2008) *Queering the Non/Human*, Burlington, VT: Ashgate.

Gilebbi, M. (2014) "Animal Metaphors, Biopolitics, and the Animal Question: Mario Luzi, Giorgio Agamben, and the Human–Animal Divide," in D. Amberson and E. Past (eds.), *Thinking Italian Animals: Human and Posthuman in Modern Italian Literature and Film*, New York: Palgrave Macmillan, 93–110.

Gillespie, K. (2018) *The Cow with Ear Tag #1389*, Chicago, IL: University of Chicago Press.

Gilligan, C. (1982) *In a Different Voice: Psychological Theory and Women's Development*, Cambridge, MA: Harvard University Press.

Giraud, E. (2013) "'Beasts of Burden': Productive Tensions between Haraway and Radical Animal Rights Activism," *Culture, Theory, and Critique*, 54: 102–120.

Glasser, C. L. (2015) "Beyond Intersectionality to Total Liberation," in Lisa Kemmerer (ed.), *Animals and the Environment: Advocacy, Activism, and the Quest for Common Ground*, New York: Routledge, 41–49.

Glock, H.-J. (2012) "The Anthropological Difference: What Can Philosophers Do to Identify the Differences between Human and Non-Human Animals?," *Royal Institute of Philosophy Supplement, 70*: 105–131.

Gold, K. C. and Watson, L. M. (2018) "In Memorium: Koko, a Remarkable Gorilla," *American Journal of Primatology, 80*: 1–3.

Goldstein, B., Moses, R., Sammons N. et al. (2017) "Potential to Curb the Environmental Burdens of American Beef Consumption Using a Novel Plant-Based Beef Substitute," *PLoS ONE, 12*: 1–17.

Goodall, J. (1986) *The Chimpanzees of Gombe: Patterns of Behavior*, Cambridge, MA: Harvard University Press.

Goodall, J. (1990) *Through a Window: My Thirty Years with the Chimpanzees of Gombe*, New York: Mariner.

Grandin, T. (ed.) (2015) *Improving Animal Welfare: A Practical Approach*, Wallingford, UK: CABI.

Gray, J. (2017) *Zoo Ethics: The Challenges of Compassionate Conservation*, Ithaca: CSIRO.

Grebowicz, M. (2010) "When Species Meat: Confronting Bestiality Pornography," *Humanimalia, 1*: 1–17.

Grebowicz, M. (2015) *The National Park to Come*, Stanford, CA: Stanford University Press.

Grebowicz, M. (2017) "Orca Intimacies and Environmental Slow Death: Earthling Ethics for a Claustrophobic World," in S. MacGregor (ed.), *Routledge Handbook of Gender and Environment*, London: Routledge, 115–130.

Grebowicz, M. and Merrick, H. (2013) *Beyond the Cyborg: Adventures with Donna Haraway; with a Seed Bag by Donna Haraway*, New York: Columbia University Press.

Greenhough, B. and Roe, E. (2018). "Exploring the Role of Animal Technologists in Implementing the 3Rs: An Ethnographic Investigation of the UK University Sector," *Science, Technology, & Human Values, 43*: 694–722.

Greek, R. and Shanks, N. (2009) *FAQs about the Use of Animals in Science: A Handbook for the Scientifically Perplexed*, Lanham, MD: University Press of America.

Gregg, J. (2013) *Are Dolphins Really Smart?: The Mammal behind the Myth*, New York: Oxford University Press.

Grier, K. C. (2006) *Pets in America: A History*, Chapel Hill, NC: University of North Carolina Press.

Griffin, D. R. (1976) *The Question of Animal Awareness*, New York: Rockefeller University Press.

Griffin, D. R. (1984) *Animal Thinking*, Cambridge, MA: Harvard University Press.

Griffin, D. R. (1992) *Animal Minds*, Chicago, IL: University of Chicago Press.

Gröning, J. and Hochkirch, A. (2008) "Reproductive Interference between Animal Species," *The Quarterly Review of Biology, 83*: 257–282.

Gruen, L. (1993) "Dismantling Oppression: An Analysis of the Connection between Women and Animals," in G. Gaard (ed.), *Ecofeminism: Women, Animals, Nature*, Philadelphia, PA: Temple University Press, 60–90.

Gruen, L. (2014a) "Dignity, Captivity, and an Ethics of Sight," in L. Gruen (ed.), *The Ethics of Captivity*, New York: Oxford University Press, 231–247.

Gruen, L. (ed.) (2014b) *The Ethics of Captivity*, New York: Oxford University Press.

Gruen, L. (2015) *Entangled Empathy: An Alternative Ethic for Our Relationships with Animals*, Brooklyn, NY: Lantern.

Gruen, L. and Jones, R. C. (2015) "Veganism as an Aspiration," in B. Bramble and B. Fischer (eds.), *The Moral Complexities of Eating Meat*, New York: Oxford University Press, 153–171.

Gruen, L. and Weil, K. (2010) "Teaching Difference: Sex, Gender, Species," in M. DeMello (ed.), *Teaching the Animal: Human Animal Studies across the Disciplines*, New York: Lantern, 127–143.

Grusin, R. (ed.) (2015) *The Nonhuman Turn*, Minneapolis: University of Minnesota Press.

Guither, H. D. (1998) *Animal Rights: History and Scope of a Radical Social Movement*, Carbondale, IL: Southern Illinois University Press.

Gusset, M. and Dick, G. (2011) "The Global Reach of Zoos and Aquariums in Visitor Numbers and Conservation Expenditures," *Zoo Biology, 30*: 566–569.

Halperin, D. (1995) *Saint Foucault: Towards a Gay Hagiography*, New York: Oxford University Press.

Hamilton, C. (2016) "Sex, Work, Meat: The Feminist Politics of Veganism," *Feminist Review, 114*: 112–129.

Hamilton, L. A. and Taylor, N. (2013) *Animals at Work: Identity, Politics, and Culture in Work with Animals*, Boston, MA: Brill.

Hancock, A.-M. (2016) *Intersectionality: An Intellectual History*, New York: Oxford University Press.

Hansen, S. (2011) "Infancy, Animality and the Limits of Language in the Work of Giorgio Agamben," *Journal for Critical Animal Studies, 9*: 167–181.

Haraway, D. J. (1985) "A Manifesto for Cyborgs: Science, Technology, and Socialist Feminism in the 1980s," *Socialist Review, 15*: 65–107.

Haraway, D. J. (2003) *Companion Species Manifesto*, Chicago, IL: Prickly Paradigm Press.

Haraway, D. J. (2008) *When Species Meet*, Minneapolis: Minnesota University Press.

Haraway, D. J. (2016a) *Manifestly Haraway*, Minneapolis: University of Minnesota Press.

Haraway, D. J. (2016b) *Staying with the Trouble: Making Kin in the Chthulucene*, Minneapolis: University of Minnesota Press.

Haraway, D. J. and Wolfe, C. (2016) "Companions in Conversation," in D. J. Haraway, *Manifestly Haraway*, Minneapolis: University of Minnesota Press, 199–296.

Hardt, M. and Negri, A. (2000) *Empire*, Cambridge: Harvard University Press.

Hargrove, E. C. (ed.) (1992) *The Animal Rights/Environmental Ethics Debate: The Environmental Perspective*, Albany: State University of New York Press.

Harper, A. B. (2011) "Connections: Speciesism, Racism, and Whiteness as the Norm," in Lisa Kemmerer (ed.), *Sister Species: Woman, Animals, and Social Justice*, Champaign, IL: University of Illinois Press, 72–78.

Harper, A. B. (ed.) (2010) *Sistah Vegan: Black Female Vegans Speak on Food, Identity, Health, and Society*, New York: Lantern.

Haslam, N. (2006) "Dehumanization: An Integrative Review," *Personality and Social Psychology Review, 10*: 252–264.

Haslam, N. and Loughnan, S. (2014) "Dehumanization and Infrahumanization," *Annual Review of Psychology, 65*: 399–423.

Hauser, M. D., Chomsky, N., and Fitch, W. T. (2002) "The Faculty of Language: What Is It, Who Has It, and How Did It Evolve?," *Science, 298*: 1569–1579.

Hauskeller, M. (2017) "How to Become a Post-Dog: Animals in Transhumanism," *Between the Species, 20*: 25–37.

Hayward, E. and Weinstein, J. (2015) "Introduction: Tranimalities in the Age of Trans* Life," *TSQ: Transgender Studies Quarterly, 2*: 195–208.

Hearne, V. (1986) *Adam's Task: Calling Animals by Name*, New York: Knopf.

Hediger, R. (2013a) "Animals and War: Introduction," in R. Hediger (ed.), *Animals and War: Studies of Europe and North America*, Netherlands: Brill, 1–28.

Hediger, R. (ed.) (2013b) *Animals and War: Studies of Europe and North America*, Netherlands: Brill.

Heidegger, M. (1962) *Being and Time*, tr. J. Macquarrie and Edward Robinson, New York: Harper and Row.

Heidegger, M. (1995) *The Fundamental Concepts of Metaphysics: World, Finitude, Solitude*, tr. W. McNeill and N. Walker, Bloomington: Indiana University Press.

Heinrich, B. (2012) *Life Everlasting: The Animal Way of Death*, Boston: Houghton Mifflin Harcourt.

Heise, U. (2016) *Imagining Extinction: The Cultural Meanings of Endangered Species*, Chicago, IL: University of Chicago Press.

Herzog, H. (2010). *Some We Love, Some We Hate, Some We Eat: Why It Is So Hard to Think Straight About Animals*, New York: Harper.

Hill, A. P. and Hadly, E. A. (2018) "Rethinking 'Native' in the Anthropocene," *Frontiers in Earth Science, 6*: 96.

Hillix, W. A. and Rumbaugh, D. A. (2004) *Animal Bodies, Human Minds: Ape, Dolphin, and Parrot Language Skills*, New York: Kluwer.

Hird, M. (2006) "Animal Transsex," *Australian Feminist Studies, 21*: 35–48.

Hodder, K. J. and Bullock, J. M. (2009) "Really Wild? Naturalistic Grazing in Modern Landscapes," *British Wildlife, 20*: 37–43.

Holland, N. J. (2018) *Heidegger and the Problem of Consciousness*, Bloomington: Indiana University Press.

Hosey, G. R., Melfi, V., and Pankhurst, S. (2013) *Zoo Animals: Behaviour, Management and Welfare*, Oxford: Oxford University Press.

Hribal, J. (2003) "'Animals are Part of the Working Class': A Challenge to Labor History," *Labor History, 44*: 435–453.

Hribal, J. (2007) "Animals, Agency, and Class: Writing the History of Animals from Below," *Human Ecology Review, 14*: 101–112.

Hribal, J. (2010) *Fear of the Animal Planet: The Hidden History of Animal Resistance*, Oakland, CA: AK Press.

Hribar, C. (2010) *Understanding Concentrated Animal Feeding Operations and Their Impact on Communities*, Bowling Green, KY: National Association of Local Boards of Health.

Hubrecht, R. C. (2014) *The Welfare of Animals Used in Research: Practice and Ethics*, Chichester, West Sussex: Wiley Blackwell.

Huggan, G. and Tiffin, H. (2010) *Postcolonial Ecocriticism: Literature, Animals, Environment*, New York: Routledge.

Human Animal Research Network Editorial Collective (eds.) (2015) *Animals in the Anthropocene: Critical Perspectives on Non-Human Futures*, Sydney: Sydney University Press.

Hursthouse, R. (2000) *Ethics, Humans, and Other Animals: An Introduction with Readings*, London: Routledge.

Hutto, J. (2014) *Touching the Wild: Living with the Mule Deer of Deadman Gulch*, New York: Skyhorse.

Imhoff, D. (2010) "Introduction," in D. Imhoff (ed.), *The CAFO Reader: The Tragedy of Industrial Animal Factories*, Berkeley: University of California Press, iii–viii.

IPCC (Intergovernmental Panel on Climate Change) (2014) *Climate Change 2014: Impacts, Adaptation, and Vulnerability, Part A*, Cambridge: Cambridge University Press.

Irvine, L. (2017) "Animal Sheltering," in L. Kalof (ed.), *The Oxford Handbook of Animal Studies*, New York: Oxford University Press, 98–112.

Iveson, R. (2012) "Domestic Scenes and Species Trouble: On Judith Butler and Other Animals," *Journal for Critical Animal Studies, 10*: 20–40.

Jackson, Z. I. (2013) "Animal: New Directions in the Theorization of Race and Posthumanism," *Feminist Studies, 39*: 669–685.

Jamieson, D. (1985) "Against Zoos," in P. Singer (ed.), *In Defense of Animals*, New York: Basil Blackwell, 108–117.

Jamieson, D. (1999) *Singer and His Critics*, Malden, MA: Blackwell.

Jenkins, S. (2012) "Returning the Ethical and Political to Animal Studies," *Hypatia, 27*: 504–510.

Jensen, P. (2002) *The Ethology of Domestic Animals*, New York: CABI.

Jepson, P. (2016) "A Rewilding Agenda for Europe: Creating a Network of Experimental Reserves," *Ecography, 39*: 169–181.

Johnson, J. T. and Larsen, S. C. (2017) *Being Together in Place: Indigenous Coexistence in a More Than Human World*, Minneapolis: University of Minnesota Press.

Johnson, L. (2017) *Race Matters, Animal Matters: Fugitive Humanism in African America, 1840–1930*, New York: Routledge.

Johnston, J. and Probyn-Rapsey, F. (eds.) (2013) *Animal Death*, NSW: Sydney University Press.

jones, pattrice (2014) *The Oxen at the Intersection: A Collision*, New York: Lantern Books.

Jørgensen, D. (2015) "Rethinking Rewilding," *Geoforum*, 65: 482–488.

Joy, M. (2011) *Why We Love Dogs, Eat Pigs, and Wear Cows: An Introduction to Carnism*, San Francisco, CA: Conari.

Kalof, L. (2007) *Looking at Animals in Human History*, London: Reaktion.

Kasperbauer, T. J. (2018) *Subhuman: The Moral Psychology of Human Attitudes to Animals*, New York: Oxford University Press.

Keeling, L. J., Rushen, J., and Duncan I. J. H. (2011) "Understanding Animal Welfare," in M. Appleby, B. Hughes, J. Mench et al. (eds.), *Animal Welfare*, Wallingford, UK: CABI, 13–26.

Kemmerer, L. (ed.) (2011) *Sister Species: Women, Animals, and Social Justice*, Urbana, IL: University of Illinois Press.

Keulartz, J. (2018) *"Rewilding,"* Oxford Research Encyclopedia of Environmental Science [Online] Available at: [http://environmentalscience.oxfordre.com/view/10.1093/acrefore/9780199389414.001.0001/acrefore-9780199389414-e-545], accessed 10 October 2018.

Kheel, M. (1996) "The Killing Game: An Ecofeminist Critique of Hunting," *Journal of the Philosophy of Sport*, 23: 30–44.

Kim, C. J. (2015) *Dangerous Crossings: Race, Species, and Nature in a Multicultural Age*, New York: Cambridge University Press.

Kim, C. J. (2018) "Abolition," in L. Gruen (ed.), *Critical Terms for Animal Studies*, Chicago, IL: University of Chicago Press, 15–32.

King, S. L., Sayigh, L. S., Wells, R. S. et al. (2013) "Vocal Copying of Individually Distinctive Signature Whistles in Bottlenose Dolphins," *Proceedings of the Royal Society B: Biological Sciences*, 280: 2–9.

King, B. (2013) *How Animals Grieve*, Chicago, IL: University of Chicago Press.

King, R. J. H. (1991) "Environmental Ethics and the Case for Hunting," *Environmental Ethics*, 13: 59–85.

Kirk, R. G. W. (2017) "Knowing Sentient Subjects: Human Experimental Technique and the Constitution of Care and Knowledge in Laboratory Animal Science," in K. Asdal, T. Druglitrö, and S. Hinchliffe (eds.), *Humans, Animals, and Biopolitics: The More-than-Human Condition*, London: Routledge, 119–135.

Kirksey, E. (ed.) (2014) *The Multispecies Salon*, Durham, NC: Duke University Press.

Kisling, V. N. (ed.) (2001) *Zoo and Aquarium History: Ancient Animal Collections to Zoological Gardens*, Boca Raton, FL: CRC Press.

Knight, A. (2011) *The Costs and Benefits of Animal Experiments*, New York: Palgrave Macmillan.

Ko, A. and Ko, S. (2017) *Aphro-ism: Essays on Pop Culture, Feminism, and Black Veganism from Two Sisters*, New York: Lantern Books.

Kolbert, E. (2014) *The Sixth Extinction: An Unnatural History*, New York: Henry Holt and Company.

Kosek, J. (2010) "Ecologies of Empire: On the New Uses of the Honeybee," *Cultural Anthropology, 25*: 650–678.

Kowalsky, N. (ed.) (2010) *Hunting—Philosophy for Everyone: In Search of the Wild Life*, Malden, MA: Wiley-Blackwell.

Kraham, S. J. (2017) "Environmental Impacts of Industrial Livestock Production," in G. Steier and K. K. Patel (eds.), *International Farm Animal, Wildlife and Food Safety Law*, Switzerland: Springer, 3–40.

Krell, D. F. (2013) *Derrida and Our Animal Others: Derrida's Final Seminar, "The Beast and the Sovereign"*, Bloomington: Indiana University Press.

Kroll, G. (2018) "Snarge," in G. Mitman, M. Armiero, and R. S. Emmett (eds.), *Future Remains: A Cabinet of Curiosities for the Anthropocene*, Chicago, IL: University Chicago Press, 81–88.

Krupenye, C., Kano, F., Hirata, S. et al. (2016) "Great Apes Anticipate that Other Individuals Will Act according to False Beliefs," *Science, 354*: 110–114.

Kurzweil, R. (2005) *The Singularity Is Near: When Humans Transcend Biology*, New York: Viking.

La Mettrie, J. O. de (1996) *Machine Man and Other Writings*, tr. A. Thomson, New York:Cambridge University Press.

Lalo, J. (1987) "The Problem of Roadkill," *American Forests, 50*: 51.

Lawlor, L. (2007) *This Is Not Sufficient: An Essay on Animality and Human Nature in Derrida*, New York: Columbia University Press.

Lawlor, L. (2008) "Following the Rats: Becoming-Animal in Deleuze and Guattari," *SubStance, 37*: 169–187.

Lemish, M. G. (1996) *War Dogs: A History of Loyalty and Heroism*, Washington, DC: Brassey's.

Lemke, T. (2011) *Biopolitics: An Advanced Introduction*, New York: New York University Press.

Lents, N. (2016) *Not So Different: Finding Human Nature in Animals*, New York: Columbia University Press.

Leopold, A. (1949) *A Sand County Almanac, and Sketches Here and There*, New York: Oxford University Press.

Lestel, D. (2014) "The Withering of Shared Life through the Loss of Biodiversity," *Social Science Information, 52*: 307–325.

Lestel, D. (2016) *Eat This Book: A Carnivore's Manifesto*, tr. Gary Steiner, New York: Columbia University Press.

Levy, N. (2003) "What (if Anything) Is Wrong with Bestiality?," *Journal of Social Philosophy, 34*: 444–456.

Lopez, B. (1998) *Apologia*, Athens, GA: University of Georgia Press.

Lorimer, J. (2015) *Wildlife in the Anthropocene: Conservation after Nature*, Minneapolis: University of Minnesota Press.

Lorimer, J., Sandom, C., Jepson, P. et al. (2015) "Rewilding: Science, Practice, and Politics," *Annual Review of Environment and Resources, 40*: 39–62.

Lovejoy, A. O. (1936) *The Great Chain of Being: A Study of the History of an Idea*, Cambridge: Harvard University Press.

Lovitz, D. (2010) *Muzzling a Movement: The Effects of Anti-Terrorism Law, Money, and Politics on Animal Activism*, New York: Lantern Books.

Luke, B. (1997) "A Critical Analysis of Hunters' Ethics," *Environmental Ethics*, 19: 25–44.

Lundblad, M. (2017) *Animalities: Literary and Cultural Studies beyond the Human*, Edinburgh: Edinburgh University Press.

Lurz, R.W. (2011) *Mindreading Animals: The Debate Over What Animals Know About Other Minds*, Cambridge, MA: MIT Press.

Lykke, N. (2009) "Non-innocent Intersections of Feminism and Environmentalism," *Women, Gender, and Research*, 18: 36–44.

MacCormack, P. (2012) *Posthuman Ethics: Embodiment and Cultural Theory*, New York: Routledge.

Mackenzie, R. (2011) "How the Politics of Inclusion/Exclusion and the Neuroscience of Dehumanization/Rehumanization Can Contribute to Animal Activists' Strategies: Bestia Sacer II," *Society and Animals*, 19: 407–424.

Maller, C. (2018) *Healthy Urban Environments: More-than-Human Theories*, New York: Routledge.

Malm, A. and Hornborg, A. (2014) "The Geology of Mankind? A Critique of the Anthropocene Narrative," *The Anthropocene Review*, 1: 62–69.

Marceau, J. (2019) *Beyond Cages: Animal Law and Criminal Punishment*, New York: Cambridge University Press.

Massumi, B. (2014) *What Animals Teach Us about Politics*, Durham: Duke University Press.

Matsuoka, A. and Sorenson, J. (eds.) (2018) *Critical Animal Studies: Towards Trans-Species Social Justice*, New York: Rowman & Littlefield International.

McArthur, J.-A. (2013) *We Animals*, New York: Lantern Books.

McArthur, J.-A. (2016) "Who Is It Acceptable to Grieve?," in M. DeMello (ed.), *Mourning Animals: Rituals and Practices Surrounding Animal Death*, East Lansing, MI: Michigan State University Press, 201–204.

McFarland, S. E. and Hediger, R. (2009) *Animals and Agency: An Interdisciplinary Exploration*, Leiden: Brill.

McKay, R. (2018) "A Vegan Form of Life," in E. Quinn and B. Westwood (eds.), *Thinking Veganism in Literature and Culture: Towards a Vegan Theory*, New York: Palgrave Macmillan, 249–272.

McMullen, S. (2015) "Is Capitalism to Blame? Animal Lives in the Marketplace," *Journal of Animal Ethics*, 5: 126–134.

McMullen, S. (2016) *Animals and the Economy*, Basingstoke: Palgrave Macmillan.

Michelfelder, D. (2003) "Valuing Wildlife Populations in Urban Environments," *Journal of Social Philosophy*, 34: 79–90.

Miletski, H. (2005) "A History of Bestiality," in A. M. Beetz and A. L. Podberscek (eds.), *Bestiality and Zoophilia: Sexual Relations with Animals*, West Lafayette, IN: Purdue University Press.

Miller, J. (2012) "In Vitro Meat: Power, Authenticity, and Vegetarianism," *Journal for Critical Animal Studies*, *10*: 41–63.

Minteer, B. A., Maienschein, J., and Collins, J. P. (eds.) (2018) *The Ark and Beyond: The Evolution of Zoo and Aquarium Conservation*, Chicago, IL: University of Chicago Press.

Mitchell, R. W., Thompson, N. S., and Miles, H. L. (eds.) (1997) *Anthropomorphism, Anecdotes, and Animals*, Albany: State University of New York Press.

Monamy, V. (2017) *Animal Experimentation: A Guide to the Issues*, New York: Cambridge University Press.

Monbiot, G. (2014). *Feral: Searching for Enchantment on the Frontiers of Rewilding*, New York: Penguin.

More, M. and Vita-More, N. (eds.) (2013) *The Transhumanist Reader: Classical and Contemporary Essays on the Science, Technology, and Philosophy of the Human Future*, Oxford: Wiley-Blackwell.

Morin, K. M. (2018) *Carceral Space, Prisoners, and Animals*, New York: Routledge.

Moses, A. and Tomaselli, P. (2017) "Industrial Animal Agriculture in the United States: Concentrated Animal Feeding Operations (CAFOs)," in G. Steier and K. K. Patel (eds.), *International Farm Animal, Wildlife and Food Safety Law*, Switzerland: Springer, 185–214.

Moss, C. (1988) *Elephant Memories: Thirteen Years in the Life of an Elephant Family*, New York: William Morrow and Company.

Mullarkey, J. (2013) "Animal Spirits: Philosomorphism and the Background Revolts of Cinema," *Angelaki*, *18*: 11–29.

Naas, M. (2015) *The End of the World and Other Teachable Moments: Jacques Derrida's Final Seminar*, New York: Fordham University Press.

Nagel, T. (1974) "What Is It Like to Be a Bat?," *The Philosophical Review*, *83*: 435–450.

Nagy, K. and Johnson, P. D. (eds.) (2013) *Trash Animals: How We Live with Nature's Filthy, Feral, Invasive, and Unwanted Species*, Minneapolis: University of Minnesota Press.

Nast, H. J. (2006) "Critical Pet Studies?," *Antipode*, *38*: 894–906.

Nayar, P. K. (2014) *Posthumanism*, Malden, MA: Polity.

Neo, H. and Emel, J. (2017) *Geographies of Meat: Politics, Economy and Culture*, New York: Routledge.

Newkirk, I. (2000) *Free the Animals: The Amazing True Story of the Animal Liberation Front*, New York: Lantern Books.

Newkirk, I. (2019) *"Why We Euthanize,"* [Online] Available at: [https://www.peta.org/blog/euthanize/], accessed 10 August 2019.

Nibert, D. (2002) *Animal Rights/Human Rights: Entanglements of Oppression and Liberation*, Lanham, MD: Rowman & Littlefield.

Nibert, D. (2013) *Animal Oppression and Human Violence: Domesecration, Capitalism, and Global Conflict*, New York: Columbia University Press.

Nibert, D. (ed.) (2017) *Animal Oppression and Capitalism*, Santa Barbara, CA: Praeger.

Nocella, A. J., Bentley, J. K. C., and Duncan, J. M. (eds.) (2012) *Earth, Animal, and Disability Liberation: The Rise of the Eco-Ability Movement*, New York: Peter Lang.

Nocella, A. J., George, A. E., and Lupina, J. (eds.) (2019) *Animals, Disability, and the End of Capitalism: Voices from the Eco-Ability Movement*, New York: Peter Lang.

Nocella, A. J., Sorenson, J., and Soch, K. (eds.) (2014) *Defining Critical Animal Studies: An Intersectional Social Justice Approach for Liberation*, New York: Peter Lang.

Nocella, A. J., White, R., and Cudworth, E. (eds.) (2015) *Anarchism and Animal Liberation: Essays on Complementary Elements of Total Liberation*, Jefferson, NC: McFarland.

Noddings, N. (1984) *Caring: A Feminine Approach to Ethics and Moral Education*, Berkeley: University of California Press.

Nordgren, A. (2012) "Ethical Issues in Mitigation of Climate Change: The Option of Reduced Meat Production and Consumption," *Journal of Agricultural and Environmental Ethics*, 25: 563–584.

Noske, B. (1989) *Humans and Other Animals: Beyond the Boundaries of Anthropology*, London: Pluto.

Nussbaum, M. (2006) *Frontiers of Justice: Disability, Nationality, Species Membership*, Cambridge, MA: Harvard University Press.

OECD/FAO (Organization for Economic Cooperation and Development/Food and Agriculture Organization of the United Nations) (2018), *OECD-FAO Agricultural Outlook 2018–2027*, Rome: OECD.

Ohrem, D. and Bartosch, R. (2017) *Beyond the Human–Animal Divide: Creaturely Lives in Literature and Culture*, Basingstoke: Palgrave Macmillan.

Oksanen, M. and Siipi, H. (eds.) (2014) *The Ethics of Animal Re-creation and Modification: Reviving, Rewilding, Restoring*, New York: Palgrave Macmillan.

Oliver, K. (2009) *Animal Lessons: How They Teach Us to Be Human*, New York: Columbia University Press.

Oppermann, S. and Iovino, S. (eds.) (2017) *Environmental Humanities: Voices from the Anthropocene*, New York: Rowman & Littlefield International.

Orlans, F. B. (1993) *In the Name of Science: Issues in Responsible Animal Experimentation*, New York: Oxford University Press.

Orozco, L. (2013) *Theater and Animals*, Basingstoke, Hampshire: Palgrave Macmillan.

Ortega y Gasset, J. (1985) *Meditations on Hunting*, tr. Howard B. Wescott, New York: Scribners.

Otomo, Y. and Mussawir, E. (eds.) (2013) *Law and the Question of the Animal: A Critical Jurisprudence*, New York: Routledge.

Overall, C. (ed.) (2017) *Pets and People: The Ethics of Our Relationships with Companion Animals*, New York: Oxford University Press.

Pachirat, T. (2018) "Sanctuary," in L. Gruen (ed.), *Critical Terms for Animal Studies*, Chicago, IL: University of Chicago Press, 337–355.

Palmer, C. and Sandøe, P. (2018) "Welfare," in L. Gruen (ed.), *Critical Terms for Animal Studies*, Chicago, IL: University of Chicago Press, 424–438.

Park, M. (2006) "Opening Cages, Opening Eyes: An Investigation and Open Rescue at an Egg Factory Farm," in P. Singer (ed.), *In Defense of Animals*, Malden, MA: Blackwell, 174–180.

Parker, S. T., Mitchell, R. W., and Boccia, M. L. (1994) *Self-Awareness in Animals and Humans: Developmental Perspectives*, New York: Cambridge University Press.

Parreñas, J. S. (2018) *Decolonizing Extinction: The Work of Care in Orangutan Rehabilitation*, Durham, NC: Duke University Press.

Paul, E. S. (2000) "Empathy with Animals and with Humans: Are They Linked?" *Anthrozoos, 13:* 194–202.

Pearce, D. (1995). *The Hedonistic Imperative*. [Online] Available at: [https://www.hedweb.com/hedab.htm], accessed 10 October 2018.

Pearson, C. (2015) "Beyond 'Resistance': Rethinking Nonhuman Agency for a 'More-than-Human' World," *European Review of History: Revue européenne d'histoire, 22:* 709–725.

Pellow, D. N. (2014) *Total Liberation: The Power and Promise of Animal Rights and the Radical Earth Movement*, Minneapolis: University of Minnesota Press.

Pepperberg, I. (1999) *The Alex Studies: Cognitive and Communicative Abilities of Grey Parrots*, Cambridge, MA: Harvard University Press.

Pepperberg, I. (2009) *Alex and Me: How a Scientist and a Parrot Discovered a Hidden World of Animal Intelligence—and Formed a Deep Bond in the Process*, New York: Harper.

PETA (People for the Ethical Treatment of Animals) (2012) *"Video: Goats Hacked Apart in Military Training,"* [Online] Available at: [https://www.peta.org/blog/leaked-video-shows-goats-hacked-apart-military-training/], accessed 6 September 2019.

Peterson, C. (2013) *Bestial Traces: Race, Sexuality, Animality*, New York: Fordham University Press.

Phelps, N. (2007) *The Longest Struggle: Animal Advocacy from Pythagoras to PETA*, New York: Lantern Books.

Philo, C. and Wilbert, C. (eds.) (2000) *Animal Spaces, Beastly Places: New Geographies of Human–Animal Relations*, London: Routledge.

Piehl, V. (2017) "Making Pig Research Biographies: Names and Numbers," in K. Asdal, T. Druglitrö and S. Hinchliffe (eds.), *Humans, Animals, and Biopolitics: The More-than-Human Condition*, London: Routledge, 48–65.

Pilsch, A. (2017) *Transhumanism: Evolutionary Futurism and the Human Technologies of Utopia*, Minneapolis: University of Minnesota Press.

Pinker, S. (1994) *The Language Instinct: How The Mind Creates Language*, New York: William Morrow.

Plumwood, V. (1996) "Being Prey," *Terra Nova, 1:* 32–44.

Plumwood, V. (2002) *Environmental Culture: The Ecological Crisis of Reason*, New York: Routledge.

Plumwood, V. (2012) *The Eye of the Crocodile*, Canberra: Australian National University E Press.

Polish, J. (2016) "Decolonizing Veganism: On Resisting Vegan Whiteness and Racism," in J. Castricano and R. R. Simonsen (eds.), *Critical Perspectives on Veganism*, Cham: Palgrave Macmillan.

Porcher, J. (2017a) "Animal Work," in L. Kalof (ed.), *The Oxford Handbook of Animal Studies*, New York: Oxford University Press, 302–318.

Porcher, J. (2017b) *The Ethics of Animal Labor: A Collaborative Utopia*, New York: Palgrave Macmillan.

Potts, A. (ed.) (2016) *Meat Culture*, Leiden: Brill.

Preece, R. (2011) *Animals and Nature: Cultural Myths, Cultural Realities*, Vancouver, BC: UBC Press.

Preece, R. and Chamberlain, L. (1993) *Animal Welfare and Human Values*, Waterloo, ON: Wilfrid Laurier University Press.

Premack, D. and Woodruff, G. (1978) "Does the Chimpanzee Have a Theory of Mind?," *Behavioral and Brain Sciences*, 1: 515–526.

Prior, J. and Ward, K. J. (2016) "Rethinking Rewilding: A Response to Jørgensen," *Geoforum*, 69: 132–135.

Puar, J. (2011) "'I Would rather Be a Cyborg than a Goddess': Intersectionality, Assemblage, and Affective Politics," *Transversal* [Online] Available at: [http://eipcp.net/transversal/0811/puar/en], accessed 6 September 2019.

Pugliese, J. (2013) *State Violence and the Execution of Law: Biopolitical Caesurae of Torture, Black Sites, Drones*, New York: Routledge.

Pyyhtinen, O. (2016) *More-than-Human Sociology: A New Sociological Imagination*, New York: Palgrave Macmillan.

Raber, K. and Mattfeld, M. (eds.) (2017) *Performing Animals: History, Agency, Theater*, University Park, PA: Penn State University Press.

Rachels, J. (1991) *Created from Animals: The Moral Implications of Darwinism*, New York: Oxford University Press.

Radick, G. (2007) *The Simian Tongue: The Long Debate about Animal Language*, Chicago, IL: University of Chicago Press.

Rasanen, T. and Syrjamaa, T. (eds.) (2017) *Shared Lives of Humans and Animals: Animal Agency in the Global North*, New York: Routledge.

Redmalm, D. (2015) "Pet Grief: When is Non-Human Life Grievable?," *The Sociological Review*, 63: 19–35.

Reese, L. A. (2018) *Strategies for Successful Animal Shelters*, San Diego, CA: Academic Press.

Regan, T. (1983) *The Case for Animal Rights*, Berkeley: University of California Press.

Regan, T. (1995) "Are Zoos Morally Defensible?," in B. G. Norton, M. Hutchins, E. F. Stevens et al. (eds.), *Ethics on the Ark: Zoos, Animal Welfare, and Wildlife Conservation*, Washington, DC: Smithsonian Institution Press.

Regan, T. (2001) *Defending Animal Rights*, Chicago: University of Illinois Press.

Renehan, R. (1981) "The Greek Anthropocentric View of Man," *Harvard Studies in Classical Philology*, 85: 239–259.

Reo, N. J. and Whyte, K. P. (2012) "Hunting and Morality as Elements of Traditional Ecological Knowledge," *Human Ecology*, 40: 15–27.

Ritchie, H. and Roser, M. (2019) *"Meat and Dairy Production,"* Our World in Data [Online] Available at: [https://ourworldindata.org/meat-production], accessed 10 September 2019.

Rivas, J. and Burghardt, G. M. (2002) "Crotalomorphism: A Metaphor for Understanding Anthropomorphism by Omission," in M. Bekoff, C. Allen, and G. M. Burghardt (eds.), *The Cognitive Animal: Empirical and Theoretical Perspectives on Animal Cognition*, Cambridge, MA: MIT Press, 9–18.

Rivera, L. (2014) "Coercion and Captivity," in L. Gruen (ed.), *The Ethics of Captivity*, New York: Oxford University Press, 248–270.

Robbins, J. (2010) *The Food Revolution: How Your Diet Can Help Save Your Life and Our World*, San Francisco: Conari Press.

Robinson, M. (2014) "Animal Personhood in Mi'kmaq Perspective," *Societies*, 4: 672–688.

Rodd, R. (1990) *Biology, Ethics, and Animals*, New York: Oxford University Press.

Rodman, J. (1980) "Paradigm Change in Political Science: An Ecological Perspective," *American Behavioral Scientist*, 24: 49–78.

Rollin, B. (1995) *Farm Animal Welfare: Social, Bioethical, and Research Issues*, Ames: Iowa State University Press.

Rollin, B. (2010) "Farm Factories: The End of Animal Husbandry," in D. Imhoff (ed.), *The CAFO Reader: The Tragedy of Industrial Animal Factories*, Berkeley: University of California Press, 6–14.

Rollin, B. (2017) "The Ethics of Animal Research: Theory and Practice," in L. Kalof (ed.), *The Oxford Handbook of Animal Studies*, New York: Oxford University Press, 345–363.

Rose, D. B., van Dooren, T., and Chrulew. M. (eds.) (2017) *Extinction Studies: Stories of Time, Death and Generations*, New York: Columbia University Press.

Rossini, M. (2006) *"To the Dogs: Companion Speciesism and the New Feminist Materialism,"* Kritikos 3 [Online] Available at: [http://intertheory.org/Rossini], accessed 10 September 2018.

Rothfels, N. (2002) *Savages and Beasts: The Birth of the Modern Zoo*, Baltimore, MD: Johns Hopkins University Press.

Roughgarden, J. (2004) *Evolution's Rainbow: Diversity, Gender, and Sexuality in Nature and People*, Berkeley: University of California Press.

Rowlands, M. (2002) *Animals like Us*, New York: Verso.

Rowlands, M. (2012) *Can Animals Be Moral?* New York: Oxford University Press.

Ruddiman, W. F. (2003) "The Anthropogenic Greenhouse Era Began Thousands of Years Ago," *Climatic Change*, 61: 261–293.

Rudy, K. (2011) *Loving Animals: Toward a New Animal Advocacy*, Minneapolis: University of Minnesota Press.

Rudy, K. (2012) "LGBTQ … Z?," *Hypatia*, 27: 601–615.

Rudy, K. (2012) "Locavores, Feminism, and the Question of Meat," *The Journal of American Culture*, 35: 26–36.

Ryder, R. D. (1971) "Experiments on Animals," in S. Godlovitch, R. Godlovitch, and J. Harris (eds.), *Animals, Men, and Morals: An Enquiry into the Maltreatment of Nonhumans*, London: Victor Gollancz.

Safina, C. (2018) "Where Are Zoos Going—or Are They Gone?," *Journal of Applied Animal Welfare Science, 21*: 4–11.

Sagoff, M. (2005) "Do Non-Native Species Threaten the Natural Environment?," *Journal of Agricultural and Environmental Ethics, 18*: 215–236.

Salt, H. S. (1894) *Animals' Rights: Considered in Relation to Social Progress*, New York: Macmillan.

Sandøe, P., Corr, S., and Palmer, C. (2016) *Companion Animal Ethics*, Ames, IA: John Wiley & Sons.

Sanz, C. M., Call, J., and Boesch, C. (eds.) (2013) *Tool Use in Animals: Cognition and Ecology*, New York: Cambridge University Press.

Scarborough, P., Appleby, P., Mizdrak, A. et al. (2014) "Dietary Greenhouse Gas Emissions of Meat-eaters, Fish-eaters, Vegetarians, and Vegans in the UK," *Climatic Change, 125*: 1–14.

Schaffner, J. E. (2011) *An Introduction to Animals and the Law*, New York: Palgrave Macmillan.

Schaler, J. A. (ed.) (2009) *Peter Singer Under Fire: The Moral Iconoclast Faces His Critics*, Chicago, IL: Open Court.

Schlottmann, C. and Sebo, J. (2018) *Food, Animals, and the Environment: An Ethical Approach*, New York: Routledge.

Schneider, J. (2005) *Donna Haraway: Live Theory*, New York: Continuum, 2005.

Schneider, S. H. and Root, T. L. (eds.) (2001) *Wildlife Responses to Climate Change: North American Case Studies*, Washington, DC: Island Press.

Science Magazine (2016) *"Humans Aren't the Only Great Apes that Can 'Read Minds,'"* [Online] Available at: [https://www.youtube.com/watch?v=1s0dO_h7q7Q], accessed 15 August 2019.

Scott, D. and Wynter, S. (2000) "The Re-Enchantment of Humanism: An Interview with Sylvia Wynter," *Small Axe, 8*: 119–207.

Seiler. A. and Helldin, J.-O. (2006) "Mortality in Wildlife Due to Transportation," in J. Davenport and J. L. Davenport (eds.), *The Ecology of Transportation: Managing Mobility for the Environment*, New York: Springer, 165–190.

Serpell, J. (1986) *In the Company of Animals: A Study of Human–Animal Relationships*, New York: Cambridge University Press.

Seshadri, K. R. (2012) *HumAnimal: Race, Law, Language*, Minneapolis: University of Minnesota Press.

Seyfarth, R. M., Cheney, D. L., and Marler, P. (1980) "Vervet Monkey Alarm Calls: Semantic Communication in a Free-ranging Primate," *Animal Behaviour, 28*: 1070–1094.

Shapiro, K. and DeMello, M. (2010) "The State of Human–Animal Studies," *Society & Animals, 18*: 307–318.

Shapiro, P. (2018) *Clean Meat: How Growing Meat without Animals Will Revolutionize Dinner and the World*, New York: Gallery.

Shukin, N. (2009) *Animal Capital: Rendering Life in Biopolitical Times*, Minneapolis: University of Minnesota Press.

Simberloff, D. (2010) "Invasive Species," in N. S. Sodhi and P. R. Ehrlich (eds.), *Conservation Biology for All*, New York: Oxford University Press.

Simberloff, D. (2011) "How Common Are Invasion-Induced Ecosystem Impacts?," *Biological Invasions, 13*: 1255–1268.

Simberloff, D. (2013) *Invasive Species: What Everyone Needs to Know*, New York: Oxford University Press.

Simonsen, R. R. (2012) "A Queer Vegan Manifesto," *Journal for Critical Animal Studies, 10*: 51–79.

Singer, P. (1993) *Practical Ethics*, New York: Cambridge University Press.

Singer, P. (2001a [1975]) *Animal Liberation*, 3rd ed., New York: Ecco/HarperCollins.

Singer, P. (2001b) *"Heavy Petting," Nerve* [Online] Available at: [https://www.utilitarian.net/singer/by/2001----.html], accessed 10 September 2019.

Singh, J. (2018) *Unthinking Mastery: Dehumanism and Decolonial Entanglements*, Durham, NC: Duke University Press.

Slicer, D. (2015) "More Joy," *Ethics and the Environment, 20*: 1–23.

Slobodchikoff, C. N. (2012) *Chasing Doctor Dolittle: Learning the Language of Animals*, New York: St. Martin's Press.

Slobodchikoff, C. N., Perla, B. S., and Verdolin, J. L. (2009) *Prairie Dogs: Communication and Community in an Animal Society*, Cambridge, MA: Harvard University Press.

Smith, D. L. (2013) "Indexically Yours: Why Being Human Is More Like Being Here than Like Being Water," in R. Corbey and A. Lanjouw (eds.), *The Politics of Species: Reshaping Our Relationships with Other Animals*, Cambridge: Cambridge University Press, 40–52.

Smuts, B. (2001) "Encounters with Animal Minds," *Journal of Consciousness Studies, 8*: 293–309.

Socha, K. (2013) "The 'Dreaded Comparisons' and Speciesism: Leveling the Hierarchy of Suffering," in K. Socha and S. Blum (eds.), *Confronting Animal Exploitation: Grassroots Essays on Liberation and Veganism*, Jefferson, NC: McFarland.

Soron, D. (2007) "Road Kill: Commodity Fetishism and Structural Violence," *TOPIA: Canadian Journal of Cultural Studies, 18*: 107–126.

Soulé, M. and Noss, R. (1998) "Rewilding and Biodiversity: Complementary Goals for Continental Conservation," *Wild Earth, 8*: 18–28.

Spellman, F. R. and Whiting, N. E. (2007) *Environmental Management of Concentrated Animal Feeding Operations (CAFOs)*, Boca Raton: CRC Press.

Spiegel, M. (1988) *The Dreaded Comparison: Human and Animal Slavery*, Philadelphia, PA: New Society Publishers.

Springmann, M., Godfray, C. J., Rayner, M. et al. (2016) "Analysis and Valuation of the Health and Climate Change Cobenefits of Dietary Change," *Proceedings of the National Academy of Sciences of the United States of America, 113*: 4146–4151.

Stallwood, K. (2013) "The Politics of Animal Rights Advocacy," *Relations: Beyond Anthropocentrism, 1*: 47–57.

Stanescu, J. K. (2012) "Species Trouble: Judith Butler, Mourning, and the Precarious Lives of Animals," *Hypatia, 27*: 567–582.

Stanescu, J. K. (2013) "Beyond Biopolitics: Animal Studies, Factory Farms, and the Advent of Deading Life," *PhaenEx, 8*: 135–160.

Stanescu, J. K. and Cummings, K. (2016a) "When Species Invade," in J. Stanescu and K. Cummings (eds.), *The Ethics and Rhetoric of Invasion Ecology*, Lanham, MD: Lexington, vii–xviii.

Stanescu, J. K. and Cummings, K. (eds.) (2016b) *The Ethics and Rhetoric of Invasion Ecology*, Lanham, MD: Lexington.

Stanescu, V. (2013) "Why 'Loving' Animals Is Not Enough: A Response to Kathy Rudy," *Journal of American Culture, 36*: 100–110.

Stanescu, V. (2016) "The Judas Pig: The Killing of 'Feral' Pigs on the Santa Cruz Islands, Biopolitics, and the Rise of the Post-Commodity Fetish," in J. Stanescu and K. Cummings (eds.), *The Ethics and Rhetoric of Invasion Ecology*, Lanham, MD: Lexington, 61–84.

Stark, H. and Roffe, J. (eds.) (2015) *Deleuze and the Non/Human*, New York: Palgrave Macmillan.

Steiner, G. (2005) *Anthropocentrism and Its Discontents: The Moral Status of Animals in the History of Western Philosophy*, Pittsburgh, PA: University of Pittsburgh Press.

Steiner, G. (2008) *Animals and the Moral Community: Mental Life, Moral Status, and Kinship*, New York: Columbia University Press.

Steward, H. (2009) "Animal Agency," *Inquiry, 52*: 217–231.

Still, J. (2015) *Derrida and Other Animals: The Boundaries of the Human*, Edinburgh: Edinburgh University Press.

Struthers Montford, K. (2016). "Dehumanized Denizens, Displayed Animals: Prison Tourism and the Discourse of the Zoo," *philoSOPHIA: A Journal of Continental Feminism, 6*: 73–92.

Suen, A. (2015) *The Speaking Animal: Ethics, Language, and the Human–Animal Divide*, Lanham, MD: Rowman & Littlefield.

Swan, J. A. (1995) *In Defense of Hunting*, New York: Harper Collins.

Taylor, A. (2009) *Animals and Ethics: An Overview of the Philosophical Debate*, Peterborough, ON: Broadview.

Taylor, C. (2013) "Foucault and Critical Animal Studies: Genealogies of Agricultural Power," *Philosophy Compass, 8*: 539–551.

Taylor, N. (2013) *Humans, Animals, and Society: An Introduction to Human–Animal Studies*, New York: Lantern.

Taylor, N. and Twine, R. (eds.) (2014) *The Rise of Critical Animal Studies: From the Margins to the Center*, London: Routledge.

Taylor, S. (2011) "Beasts of Burden: Disability Studies and Animal Rights," *Qui Parle: Critical Humanities and Social Sciences*, 19: 191–222.

Taylor, S. (2013) "Vegans, Freaks, and Animals: Toward a New Table Fellowship," *American Quarterly*, 65: 757–764.

Taylor, S. (2017) *Beasts of Burden: Animal and Disability Liberation*, New York: The New Press.

The Animal Studies Group (eds.) (2006) *Killing Animals*, Urbana, IL: University of Illinois Press.

The Vegan Society (n.d.) *"Definition of Veganism,"* [Online] Available at: [https://www.vegansociety.com/go-vegan/definition-veganism], accessed 12 September 2018.

Thierman S. (2010) "Apparatuses of Animality: Foucault Goes to a Slaughterhouse," *Foucault Studies*, 9: 89–110.

Thomas, S. and Shields, L. (eds.) (2012) "Special Issue: Prison and Animals," *Journal of Critical Animal Studies*, 10: 4–139.

Tinbergen, N. (1963) "On the Aims and Methods of Ethology," *Zeitschrift für Tierpsychology*, 20: 410–433.

Tischler, J. (2008) "The History of Animal Law, Part I (1972–1987)," *Stanford Journal of Animal Law and Policy*, 1: 1–49.

Tischler, J. (2012) "A Brief History of Animal Law, Part II (1985–2011)," *Stanford Journal of Animal Law and Policy*, 5: 27–77.

Todd, Z. (2018) "Refracting the State through Human–Fish Relations: Fishing, Indigenous Legal Orders, and Colonialism in North/Western Canada," *Decolonization: Indigeneity, Education & Society*, 7: 60–75.

Toledo, D., Agudelo, M. S., and Bentley, A. L. (2011) "The Shifting of Ecological Restoration Benchmarks and Their Social Impacts: Digging Deeper into Pleistocene Re-wilding," *Restoration Ecology*, 19: 564–568.

Tønnessen, M., Oma, K. A., and Rattasepp, S. (eds.) (2016) *Thinking about Animals in the Age of the Anthropocene*, Lanham, MD: Lexington.

Tønnessen. M. and Oma, K. A. (2016) "Introduction: Once upon a Time in the Anthropocene," in M. Tønnessen, K. A. Oma, and S. Rattasepp (eds.), *Thinking about Animals in the Age of the Anthropocene*, Lanham, MD: Lexington, vii–xix.

Torres, B. (2007) *Making a Killing: The Political Economy of Animal Rights*, Oakland, CA: AK Press.

Tuan, Y.-F. (1984) *Dominance and Affection: The Making of Pets*, New Haven: Yale University Press.

Tuck, E. and Yang, K. W. (2012) "Decolonization Is Not a Metaphor," *Decolonization: Indigeneity, Education & Society*, 1: 1–40.

Turner, L. (ed.) (2013) *The Animal Question in Deconstruction*, Edinburgh: Edinburgh University Press.

Twine, R. (2010) "Intersectional Disgust?—Animals and (Eco) Feminism," *Feminism and Psychology*, 20: 397–406.

Twine, R. (2012) "Revealing the 'Animal–Industrial Complex': A Concept and Method for Critical Animal Studies?" *Journal for Critical Animal Studies*, 10: 12–39.

U.S. Department of the Interior, U.S. Fish and Wildlife Service, and U.S. Department of Commerce, U.S. Census Bureau (2016) *National Survey of Fishing, Hunting, and Wildlife-Associated Recreation* [Online] Available at: [https://wsfrprograms.fws.gov/subpages/nationalsurvey/nat_survey2016.pdf] accessed 12 September 2018.

Union of Concerned Scientists (2016) *"The Planet's Temperature Is Rising,"* [Online] Available at: [https://www.ucsusa.org/global-warming/science-and-impacts/science/temperature-is-rising], accessed 28 September 2018.

van der Ree, R., Smith, D.J., and Grilo, C. (eds.) (2015) *Handbook of Road Ecology*, Hoboken, NJ: Wiley-Blackwell.

van Dooren, T (2014) *Flight Ways: Life and Loss at the Edge of Extinction*, New York: Columbia University Press.

Varner, G. (2001) "Sentientism," in D. Jamieson (ed.), *A Companion to Environmental Philosophy*, Malden, MA: Blackwell, 192–203.

Veracini, L. (2010) *Settler Colonialism: A Theoretical Overview*, New York: Palgrave Macmillan.

von Uexküll, J. (1985) "Environment and Inner World of Animals," tr. C. J. Mellor and D. Gove, in G. M. Burghardt (ed.), *Foundations of Comparative Ethology*, New York: Van Nostrand Reinhold, 222–245.

von Uexküll, J. (2010) *A Foray into the Worlds of Animals and Humans: With a Theory of Meaning*, tr. J. D. O'Neill, Minneapolis: University of Minnesota Press.

Wadiwel, D. J. (2015) *The War against Animals*, Boston, MA: Brill.

Wadiwel, D. J. (2018) "Biopolitics," in L. Gruen (ed.), *Critical Terms for Animal Studies*, Chicago, IL: University of Chicago Press, 79–98.

Waisman, S. S., Frasch, P. D., and Wagman, B. (2014) *Animal Law: Cases and Materials*, Durham, NC: Carolina Academic Press.

Walters, K. S. and Portmess, L. (eds.) (2001) *Religious Vegetarianism: From Hesiod to the Dalai Lama*, Albany: State University of New York Press.

Warkentin, T. (2012) "Must Every Animal Studies Scholar Be Vegan?," *Hypatia*, 27: 499–504.

Warren, M. A. (1997) *Moral Status: Obligations to Persons and Other Living Things*, New York: Oxford University Press.

Watts, V. (2013) "Indigenous Place-Thought and Agency Amongst Humans and Non-Humans (First Woman and Sky Woman go on a European World Tour!)," *Decolonization: Indigeneity, Education & Society*, 2: 20–34.

Weheliye, A. G. (2014) *Habeas Viscus: Racializing Assemblages, Biopolitics, and Black Feminist Theories of the Human*, Durham, NC: Duke University Press.

Weinstein, J. and Colebrook, C. (eds.) (2017) *Posthumous Life: Theorizing Beyond the Posthuman*, New York: Columbia University Press.

Weis, T. (2007) *The Global Food Economy: The Battle for the Future of Farming*, London: Zed.

Weis, T. (2013) *The Ecological Hoofprint: The Global Burden of Industrial Livestock*, London: Zed.

Weisberg, Z. (2009) "The Broken Promises of Monsters: Haraway, Animals, and the Humanist Legacy," *Journal for Critical Animal Studies*, 7: 21–61.

Wenzel, G. (1991) *Animal Rights, Human Rights: Ecology, Economy, and Ideology in the Canadian Arctic*, London: Belhaven Press.

Westling, L. (2014) "Très Bête: Evolutionary Continuity and Human Animality," *Environmental Philosophy*, 11: 1–16.

Whiten, A and Mesoudi, A. (2008) "Establishing an Experimental Science of Culture: Animal Social Diffusion Experiments," *Philosophical Transactions of the Royal Society of London B, 363*: 3477–3488.

Wilson, D. A. H. (2015) *The Welfare of Performing Animals: A Historical Perspective*, Heidelberg: Springer.

Winders, B. and Ransom, E. (eds.) (2019) *Global Meat: Social and Environmental Consequences of the Expanding Meat Industry*, Cambridge, MA: MIT Press.

Winograd, N. J. (2007) *Redemption: The Myth of Pet Overpopulation and the No Kill Revolution in America*, Los Angeles, CA: Almaden.

Winograd, N. J. and Winograd, J. J. (2017) *Welcome Home: An Animal Rights Perspective on Living with Dogs and Cats*, San Francisco, CA: Almaden.

Wise, S. M. (2000) *Rattling the Cage: Toward Legal Rights for Animals*, Cambridge, MA: Perseus.

Wise, S. M. (2005) *Drawing the Line: Science and the Case for Animal Rights*, Cambridge, MA: Perseus.

Wolch, J. (1998) "Zoopolis," in J. Wolch and J. Emel (eds.), *Animal Geographies: Places, Politics, and Identity in the Nature-Culture Borderlands*, London: Verso, 119–138.

Wolfe, C. (2010) *What Is Posthumanism?* Minneapolis: University of Minnesota Press.

Wolfe, C. (2013) *Before the Law: Humans and Other Animals in a Biopolitical Frame*, Chicago, IL: University of Chicago Press.

Wolfe, P. (1999) *Settler Colonialism and the Transformation of Anthropology: The Politics and Poetics of an Ethnographic Event*, New York: Cassell.

Womack, C. (2013) "There Is No Respectful Way to Kill an Animal," *Studies in American Indian Literatures*, 25: 11–27.

Wrenn, C. L. (2015) *A Rational Approach to Animal Rights: Extensions in Abolitionist Theory*, New York: Palgrave Macmillan.

Wright, L. (2015) *The Vegan Studies Project: Food, Animals, and Gender in the Age of Terror*, Athens: University of Georgia Press.

Wright, K. (2017) *Transdisciplinary Journeys in the Anthropocene: More-than-Human Encounters*, New York: Routledge.

Wurgaft, B. A. (2019) *Meat Planet: Artificial Flesh and the Future of Food*, Berkeley: University of California Press.

Zylinski, S. (2015) "Fun and Play in Invertebrates," *Current Biology, 25*: 10–12.

INDEX

CPSIA information can be obtained
at www.ICGtesting.com
Printed in the USA
LVHW011550250821
696067LV00017B/2271